The Joy *of* Plumbing

A guide to living the life you really, **really** want

Dedication

To all those women that got ideas 'above their station,' bucked the trends and created new ones. To those that have gone before me, and those that will live the life they really, ***really*** *want after me. I dedicate this book to you*

The Joy *of* Plumbing

A guide to living the life you really, **really** want

Hattie Hasan

The Joy of Plumbing

First published in 2011 by;
Ecademy Press
48 St Vincent Drive, St Albans, Herts, AL1 5SJ
info@ecademy-press.com
www.ecademy-press.com

Printed and Bound by; Lightning Source in the UK and USA
Set in Warnock and Gill Sans by Karen Gladwell
Cover photography and illustrations by Michael Inns

Printed on acid-free paper from managed forests. This book is printed on demand, so no copies will be remaindered or pulped.

ISBN 978-1-907722-09-7

A CIP catalogue record for this book is available from the British Library.

Contents

Acknowledgements

This is a book I was destined to write. It has lain dormant throughout my life, even before I made the decision to control my own destiny.

It could not have been possible without the support of very special people around me that I would like to thank publicly.

My mum, Huriye Hasan; without your love and strength, I would not be the person I am today.

My dad, Ali Hasan; you have unwittingly passed on your stubborn streak that I have used to succeed.

My siblings, Ali, Oz, Fatma and Sue; growing up with you all was how I shaped my world into one I wanted. You are all remarkable people.

Huriye Solkanat; You have given me the inspiration to take my message to the next generation.

Mica May; Your support and love throughout has been priceless. Your ability to help me clarify my thoughts has enabled me to go beyond. Thank You.

Thanks also, to those who believe in me and support me to go further.

Hazel Davis, Daniel Priestley, Andrew Priestley, Mike Southon, Michelle Clarke, Mindy Gibbins-Klein, Roger Hamilton. Ian Rutter, Mike Symes, Neil Davidson.

Thanks to the inventors of Twitter, FaceBook and You Tube, for making it so easy to get my message out and helping me to connect with women plumbers from around the world.

Also, to those women plumbers that talked to me about their passion for plumbing and have contributed in kind to this book. Thank you to Claire Czajkowska and Mo Pimenta for your special help.

And finally, those that have inspired me that I don't know. Hedy Lamarr, Amelia Earhart, Anne Frank, Nelson Mandela, Kate Bush.

Introduction

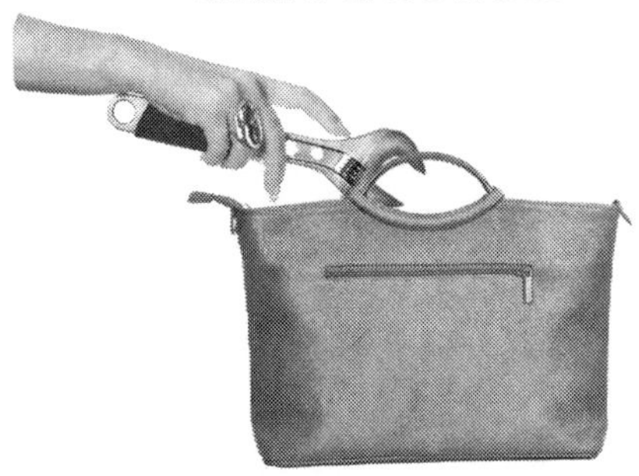

What's it all about?

Why am I writing a book called **The Joy of Plumbing**? What's so joyous about plumbing that it needs a whole book dedicated to it?

I'm going to take you on a journey. This is probably not like any other you've ever been on. Obviously, reading any book is a journey in itself and some expeditions are fun, fun, fun. Others are hard work but you still feel you have to go on them. A voyage of self-discovery is one of the most difficult and yet the most rewarding journeys you can go on.

This particular one you will compose yourself. Huh? Well, what I mean is, I will be the guide. This journey has no fixed end and no fixed beginning and the route is one that moves through space and time as a hot knife through butter.

OK, let me make sense of that.

You'll see as you go through this book, that it was no accident that I became a plumber and that after 20 happy years in the business of 'controlling water' I now want to share how liberating it was for me and encourage you to take the step further and realise that plumbing IS a joy, and that it can FREE your life.

You'll see that it was no accident that you are doing what you are doing and that you were as drawn to read this book as I was to write it.

What has all this got to do with plumbing?

I'm sure you've asked yourself on many occasions, 'Why am I doing this?' 'Why have I chosen to do something that society tells me I shouldn't?' 'Why don't I just go back to doing what I was safe doing?'

I've asked myself those questions too.

I was a very successful and 'safe' primary school teacher. What about you?

In my research for this book I asked 50 women plumbers from the UK and USA why they chose plumbing. You'll probably not be surprised to hear that it was not for the sewage and toilet jobs. (Although I always had a fascination for mushrooms and fungi as a child, being as they are nature's rot eaters). Many chose plumbing because of the variety, the day-to-day challenge, the money, the excitement of solving problems. These are needs that can be fulfilled by most jobs. Being a manager on the shop floor can give you these things but what is it about water? (Here's where I go into my passion of talking about water. I've kept it to a minimum but if you don't like water (!!) then miss out the next few paragraphs).

For one thing, water is always on the move. Water is free. Water always finds its own level. Water is what keeps us alive. Humans can survive for weeks without food, but only days without water. Water is life. The planet is two-thirds covered with the stuff and our bodies are 70% made up of it. It carries our life force and our genes into the next generation. It transports our waste and cleans our systems. The oceans and seas are clearly linked to the moon and its cycles, just as we are linked through our menstrual cycles.

A smart factory boss in Minnesota was worried about the levels of output at his packing plant. He noticed that a pattern emerged. Every month there were days of decreased production. Machines would break down, workers would be agitated, complaints would increase. He hired an analyst to confirm his observations, that there was indeed a pattern. The analyst, being a woman, felt she could better understand the workforce who were also predominantly women if she spent time with them during these unproductive periods and that is exactly what she did. It soon became clear what was causing the disruption. She realized that the monthly cycles of the female workers had synchronized. She recommended that the factory close down for the three days most effected every month. After some number crunching the factory boss decided to try it out for three months. He found that rather than losing productivity overall, it was improved for the rest of the month. It became company policy.

Do not underestimate the power of water or women. Combine the two, and there is a force to be reckoned with.

The ingenious ones who discovered that we could imitate nature and bring water into our homes, use water for heat, and use water to carry out our waste were the creators of true civilization.

The added bonus of working with metal was a clincher for me too. To be able to design a system to carry out a specific purpose. To build and create that system and then to see it working, is very satisfying indeed.

Why am I writing this book?

It has always been one of my 'things' to pass on knowledge. From an early age sharing a bedroom with my two younger sisters meant I had my very own 'classroom'. I would read to them and teach them and help them with their homework and this was very satisfying for me. I always had a fascination for wanting to know how things worked. I was always taking things apart and putting them back together again. I also did experiments at home and it was often a loud bang could be heard as I yet again blew the fuses with one of my 'brilliant' ideas.

So this was me. I loved to teach and I loved to learn and I loved to do. My entrepreneurial streak came out in high school. I had a nice little business going repairing gadgets for my friends and I earned the nickname 'Maintenance'. The 70s might have felt modern but as a girl I was not allowed to study my chosen metalwork and engineering but was instead forced towards 'safe' cookery and needlework. My interest in teaching was encouraged though and my career seemed set. I started teaching in inner-city London with some of the poorest and deprived children.

For ten years that's what I did and my water lust went unmet.

How long can you keep doing something that you were not meant to do? I loved my job, I loved the opportunity it gave me to form young minds. I loved the long breaks and the money meant I was able to take good holidays. I indulged my practical side by building sheds and doing up my own homes. But there was something missing. Something that I had buried in the passing years that nagged at me and pulled me.

Towards the end of the 80s education changed. It became more of a tick box job than a teaching job and I took the opportunity to leave. Now I could do what I wanted. I enrolled into the local college and went back to school to train to be a plumber.

In 1990 I upped sticks and left for the North, where I continued my training and looked for work. My belief was that you train, then you work with someone, then you either stay with them or you start your own business. This seemed to work for all the men on my course, but not for me. It didn't take long for me to realise that there would be no apprenticeship for me. Such comments as, 'It's a dirty job you know?' 'Are you calling up for your husband?' and 'You do know we're a plumbing firm don't you?' were common. If I wanted to live I had to work for myself.

STOPCOCKS
Woman plumber
Anything from
dripping taps
to
fitted kitchens
Also bathrooms and showers
installed
Free estimates within the
local Leeds area
Phone Hattie
Leeds 489550

In June 1990 Stopcocks was born. I created a crude little logo and advertised in the local rag.

The phone rang off the hook and in my first month I had already worked for three of the weeks and had work waiting for me in the following month. Since then I have moved home three times and in every new location Stopcocks has thrived.

Plumbing has freed my life and although this book will not tell you how to run your plumbing business, it will tell you how you are so well equipped to, why your business will succeed and how you can start earning good money for yourself straight away. If you are near a computer you may want to have a look at **www.stopcocks.co.uk** to see what I'm talking about.

Don't worry if you're on a train or in a coffee bar. There will be reminders of this address and many of the other things you can get as you read through.

This book won't tell you how to 'plumb' either. OK, so you're reading a book called **The Joy of Plumbing** and I've just told you I won't be telling you how to plumb, or how to run a business.

What on earth is the book about then?

True that my plumbing business freed my life. True that being a great plumber has done immeasurable things for my confidence. True that passing on all of that is what I intend my legacy to be. True that all I spend my days doing these days is talking to educators, media and other women plumbers about the great wave of change that is happening to the domestic construction industry in the UK.

Things are changing for the better. The customer is being heard and those that can't supply a high level of service will be flushed out. Pun intended. My own empirical research over the years has shown that what the customer wants above all is respect. This may come as a surprising fact, but the overriding reason people use female traders is because they perceive them as more trustworthy than their male counterparts. There will be more on that later and an explanation of 'bad boy' syndrome, that seemed to engulf the construction industry during the nineties and proved difficult to shift in the new millennium.

This is not a man-bashing book

Like it or not, we need men. This book is not intended to bash the guys out there. We all know a good man when we see one. I'm sure you've had experiences good and bad with a male plumber. You know when you've found a plumber you can trust you almost don't care how much he charges or how long it takes for him to get to your job. The chances are you will wait for non-urgent things because you know he will always do a great job. We also know that male plumbers have the worst reputation amongst tradesmen. Why is this? I'll be exploring that in more detail later on.

Be under no illusion though, when you start your plumbing business (read electrician, joinery or building if you must) your customers will almost not care how much you charge or how long they will have to wait for you to get round to their job. They will wait for non-urgent work, because they know you always do a great job.

Who is this book for?

In 2006 I received an email from a woman who worked as an airline hostess. Coming towards the end of her twenties she was thinking that her safety announcement days would soon be over and besides, she was wanting to be in one place for longer than four hours at a time so she contacted me. I was able to advise her on how to start changing her career and her life.

I received many other emails from nurses and admin workers and at the time of writing, from an ex army soldier. All women wanting to reinvent themselves and take control of their lives by becoming plumbers.

So to answer the question. This book is for you if you are a woman wanting to take control of your life. Whether you are just starting out or well into your working life. Age doesn't matter here. It's never too early or too late. Funny enough, it doesn't really matter if plumbing is your 'bag' or not, though I really hope it is because that way I will probably get to meet you one day. The chapters and stories in this book are just the beginning of your new life if you are a woman training to be a plumber or not.

Are you in the NOT category? Are you already plumbing and want to take your business to the next level? Are you an inspirer or an educator? Do you remember a time when you lacked confidence and how difficult the journey was to gain and grow it? How many women do you know who are allowing their lack of confidence to hold them back. When you look at your friends, do you see the potential in them? Do they see the potential in you?

How many times do we need to learn something before we've learned it for real?

We will be exploring common themes such as the three Cs; Confidence, Competence, Character and these will be cropping up as you progress on this journey of self discovery, self belief and plumbing.

It's not just about fixing pipes

For me plumbing has never been about fixing pipes. There's so much more to it than that. In researching this book I had many conversations with plumbers both male and female. Talking to old school plumbers is a joy. Together we wax lyrical about the beauty of copper and steel, and share stories of frogs in pipes and how a particular problem was solved. The form and function of a system that is pipes and water but that gives so much pleasure to the customer.

After a hard day's work I arrived home to find a message on my answering machine. It said, 'I just wanted to ring you up and say thank you for my central heating. I love it, I have never felt so snuggly and I want to say THANK YOU SO MUCH.' That kind of message makes my day.

I've been in their house for three days, I've taken up floorboards and drilled holes and charged them money. Yet three weeks later they want to express their gratitude. That is cool.

Imagine yourself now. You are at home. You have a toddler who's just become mobile. You've just put her/him in the high chair for their lunch. Just as you're about to put in the first spoonful of yummy goo you hear a creak and the high chair lilts over.

What do you do?

You drop the food and grab junior right? Then what?

OK so let's assume baby is fed and watered all safe and sound and is now napping peacefully in the next room. As you enter the kitchen you sigh at the sight of the still crooked high chair but notice a nut and bolt on the floor. You can see that it has fallen free from the high chair and that this was the cause of the mishap. Now you have some choices. Do you sit on the floor and put the chair back together or do you wait till daddy gets home?

What you do next can really alter the rest of your day.

Scenario 1. *Imagine you sit and put the chair together. You find a spanner and put the chair together. Even though you really know nothing about high chair construction you have succeeded.*
How does it make you feel?

Scenario 2. *Now imagine, you picked up the nut and bolt and put them on the side for daddy to fix when he gets home.*
How does that make you feel?

What will happen at tea-time in scenario one? I can hear you now, 'Mummy fixed the broken chair and now you can sit in your own chair that mummy fixed. Who is a clever mummy?' What a great mood you're in. How confident you feel. When your man gets in you are in high spirits because you solved the problem, fixed the chair and all is well. He is in for a treat.

What about scenario 2? You're grumpy because you can't put junior in the high chair which means you're going to have to feed her/him on your lap, which is more complex and complicated. You're annoyed because your man is not home in time to fix the chair before tea and when he gets home the poor guy gets your stress in the neck.

Forgive me if this is a 'da!' moment and I really am stating the obvious, but sometimes the obvious needs to be stated.

Think back to scenario 1. All you did was put a nut and bolt together. It's not rocket science, and yet it made you feel great. Why did you feel so confident? In scenario 2. Why did you feel so grouchy? Did it have anything to do with your husband/partner?

Now imagine a scene that is more probable (if you are already training to be a plumber). You get called to a house where the guy has no interest in DIY at all. This is evident because he tells you. 'I don't know anything about how the water works in this house, I would much rather pay you.'

You have no idea what you will be facing, all you know from his description is that there is water coming out of a pipe into the garden.

Stop there.

What do you think?

How confident do you feel?

What preparations will you make before turning up?

What tools do you think you will need?

How long will you allow for the job?

Write down what could be the possible causes, and answer the questions. **Then** turn over the page and continue.

The problem is, there's water flowing freely from a pipe into the back yard. On further inspection (which he has had no interest in doing) you discover it is from an overflow pipe attached to a toilet.

Stop there.

> **Were you right with your diagnosis?**
>
> **Do you have the know-how and the tools to make the repair?**
>
> **How confident do you feel?**

From a simple exercise like this you can easily see that from limited knowledge about the actual problem, you were able to make quite an accurate diagnosis and do a competent repair.

There is more to plumbing than fixing pipes. Building up your confidence by noticing how these small things add up not only increases your competence but also builds your character.

In my first year as a plumber I carried every single tool I owned in the back of the van just in case I needed anything. How my suspension suffered. My ambition was to carry as little as possible but always have the means to get the job done. I achieved that a couple of years later and I've made a You Tube video to show it.

Confidence has a habit of creeping all over your life. Once you have it, it moves into your personal life and your social life. It just can't help showing itself whenever you speak to anyone. Of course, there will always be a need for plumbers. You will be very popular at any party. So much so that I have dedicated a whole chapter to why your business will be so successful.

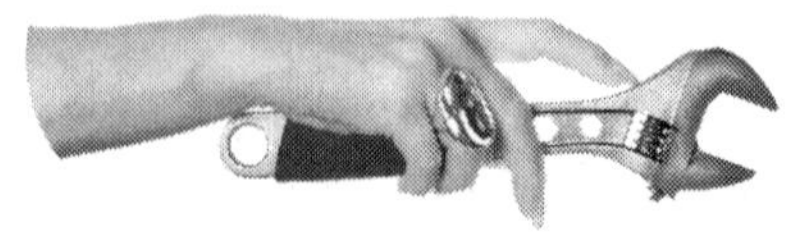

And finally...

Before we embark on this journey with no fixed start end or route there is the question of legacy. We're going to leave a legacy anyway, so why leave it to chance? Why not create an intentional legacy. My ancestry comes from a land where water is scarce. Not because of lack of rain believe it or not, but simply because the country is divided and most of the rain falls on the 'other side'. Water runs through the taps until midday and then in the summer when it is needed most it is cut off. Enough water has to be collected to last the day. This was brought home to me when I visited my ancestral lands and witnessed it with my own eyes. Here the women collect water while it is flowing and eke it out when it is turned off. I can be grateful that my extended family have the water flowing for as long as they do. No, I do not come from the developing worlds of Africa or India. My ancestry is Mediterranean. Water issues are not confined to the developing world. Just look at events over the years in New Orleans, the water shortages and hosepipe bans of the UK and the amazing work being done by women in Kenya building corrugated iron roofs to collect rainwater and saving hundreds of children's lives everyday. My belief is that there is enough water on the planet to sustain every living creature and plant.

Women have carried water on their backs for millennia to sustain their villages. As a WOMAN I feel this, even from my own ancestry. As a PLUMBER I can do something to change it. If I do not I will have missed the greatest opportunity to create the changes I want to see.

So dear reader, decide.

Are you coming?

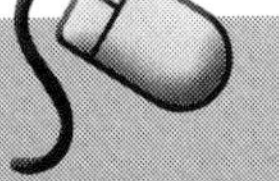

See us at **www.stopcocks.co.uk**

Watch the Stopcocks Channel at **www.youtube.com/stopcocks**

CHAPTER ONE

The Art of Reinvention

Reinventors of our Times

J.K Rowling started her working life as a researcher and bilingual secretary for Amnesty International.

Anita Roddick of Body Shop fame, started as a teacher and restaurant owner.

Margaret Thatcher, worked as a chemist in an ice-cream factory.

I started as a tobacco seller, motorcycle despatch rider, then teacher.

Women are the mistresses of reinvention. It is a rare woman that leaves school and does just the one thing until she retires. Most successful women will have reinvented themselves many times.

This book is for you if you are considering reinventing yourself. The fact that you have chosen this book shows that you are willing to look outside of the ordinary and are considering going through the reinvention process. You are in good company. Look at the biographies of successful women and you will see time and again that their success came during their third or fourth reinvention.

A woman is still more likely to be torn between career and family, she is still more likely to follow and support her husband as he climbs the career ladder to success. It is still by necessity that a woman leave her career while she gives birth and society still pressures her to stay at home while her children grow up. Even when the economic climate dictates that everyone go out to work, it is still the women who bear the guilt of leaving the children in childcare, while it is still commonly accepted that the man will go out to work.

In modern economies in the West, the overriding attitude is still that women leaving their children in childcare are 'bad mothers', whereas men doing the same are 'good providers'.

The point here is not to slate the system, although in a conversation with Richard Reed, founder of Innocent Drinks, he told me how they rate their female workers so highly that they give new mothers a bonus of £2000 to help with childcare in the first year. (I hugged him for that.)

No, the point is to show that we are constantly reinventing ourselves (and not always by choice) and it is once we have tried things and done things and learnt things that we finally settle on what we want to do to make our lives fulfilled.

We often go on reinventing ourselves when our husbands are putting their feet up ready to rest. After working hard all his life he is ready to settle down into retirement. We've nurtured our children into their independent lives and supported him through his career. Now we see this as 'our time' and grab the opportunity to do something even more fulfilling with our lives.

During my years as a teacher I witnessed many women who had been in the job for years. I nearly got trapped there myself. Teaching, particularly primary school teaching, which was my sector is a very female dominated area. Perhaps because the children are young and in some ways this can be seen as an extension of or even a substitute for being a mother. My experience here was both inspiring and revealing.

I was inspired daily by the children I taught but what was revealed to me, was in many ways, much more interesting.

A colleague had been in the profession for more than 40 years. She no doubt loved her job and had risen to the ranks of deputy head. She was beloved by the children too. As I got to know her I saw that as well as her own children she also had adopted a child, and that she was active in her community and did many charitable works.

I discovered that she was trapped in her job. She wanted to travel, and learn, and dance and sing but instead she covered up those needs (I can only assume she had inner feelings of guilt as she was of the generation where having fun was self-indulgent) and overcompensated by doing good and worthy works out in the community. As I got to know her more she also revealed that she had health difficulties, arising from these inner conflicts.

In some ways I can thank her for showing me that it is not good to ignore your dreams. I say I nearly got trapped there too because teachers are often made to feel that they can do nothing else, and as I began to break away from the profession, I certainly felt that way.

How Comfortable is Your Comfort Zone?

Have a look at where you are at right now. Let's talk about the myth that is the 'comfort zone'. That term is generally applied to doing the same old thing. 'Better the devil you know' I would encourage anyone and everyone to challenge these clichés made to keep people from changing and improving their situations.

Before the Industrial Revolution 200 years ago, communities lived in small groups. They had only the resources available to them within walking (or a bit further if they owned a cart) distance. In those communities nearly everyone was an entrepreneur. If you needed some shoes, you went to the local cobbler. For clothing, you went to the local seamstress. If you wanted bread, you went to the local baker. For stone work, you saw the local stone mason.... you get the picture. When the Industrial Revolution swept across the country those who had enough money bought factories. What do you suppose happened to the small cottage industries?

Now, instead of paying high prices and waiting for your shoes, you could buy them cheaper from the factory. What about the shoe makers who made shoes for the village? They became the factory workers. Now they were working to contribute to the wealth of the factory owner, who could pay the workers a pittance, because they had no where else to go.

So let's look at those terms again. 'comfort zone' It would be considered way out of the shoe maker's 'comfort zone' to rise up against the shoe factory owner, even though their situation was far from comfortable. Individuals stood no chance against the factory owners. For them it was 'better the devil you know'

Two hundred years later and the situation is now opposite to that. Large corporations and factories have become so large that it takes them years to change to new situations. The 'devil we know' is a dinosaur that we can all outrun. Those of you that look outside your uncomfortable 'comfort zone' and leave behind the 'devil you know' will see that being in control of your own destiny is a thrill better than any roller coaster ride.

Have you ever done something just for the thrill? Have you ever challenged yourself to do something you never though you would? Funny I should use the roller coaster example as my example for a thrill. I hate roller coasters. They are my idea of the ultimate nightmare. Yet I was challenged to take a ride on one of the biggest roller coasters in Europe. I had told myself I would do it. I had psyched myself up and just to make sure I did it I asked friends to sponsor me to raise money for WaterAid. Now I was destined to jump out of my 'comfort zone' I couldn't back down. On the morning of the ride I went with my niece (adrenalin junkie and roller coaster aficionado). I stood at the bottom and looked up at the 235ft (72 m) ride. To console myself I watched the rides before me with a stop-watch. Only three minutes. I took a deep breath in. Only three minutes and then it'll be over. My niece wanted to take me to the front carriage when our turn came, but I would have rather stuck hot pins in my eyes and she agreed to sit somewhere in the middle. As we waited I held my chest so as to stop my heart from bursting through. I remembered a friend telling me that fear and excitement felt the same, and tried to take comfort from that. 'Three minutes, three minutes', my mantra went. As the ride took off and started its slow ascent I remembered to breathe and told myself that I was now in the lap of the gods. Whatever happens will happen and I have no way to do anything about it. I closed my eyes and for the next two and a half minutes I travelled at speeds of up to 87mph (mainly downhill) and I screamed at the top of my lungs until we came in to land safe and sound.

I had done it! I had faced one of my biggest fears and come out hoarse but victorious. I also raised a wad of cash.

Has that made me love the roller coaster? Well, no, but what I did on that day meant I could go on all the other rides in the park with no fear. The bigger effect though was this cliché.

'There is nothing to fear but fear itself'

Nobody ever said that reinvention would be easy. When most people are telling me that it would be too hard, I prefer to listen to those that tell me 'if anyone can do it, you can'.

Who will you be listening to? And does anyone really take advice anyway? From an early age we are programmed to explore, to learn, to push the boundaries. If I were to listen to my mother's panicked warnings I would never have learned that a stinging nettle would give me a rash, but it wouldn't kill me. It is only by experiencing that we truly learn.

It has always been necessary, to survive and thrive. Who coined the saying....

'Necessity is the mother of invention'

Let me coin my own phrase here.

'Fulfillment is the mother of reinvention.'

Perhaps it sounds and feels counterintuitive to say that. Can you say that you are fulfilled in your life? Are you making a living or living a life?

Why, when I was earning a great regular salary, with a beautiful home and car and all the 'trappings' did I decide to leave it all and reinvent myself? I had reached the end of my growth as a teacher and needed to move on to continue growing. Like a sapling that has outgrown its pot. It will not thrive and grow unless it has the room to do it. To make a bonsai tree you only need to keep the tree in a small pot and keep cutting its roots and branches. I was not content to be a bonsai even if I could be the best bonsai in the pack. I needed to grow.

So here's where you get to really look at your own life. Can you see the different phases? When did you first reinvent yourself? Did you put a dream on the back burner while you supported your partner? Did you leave a promising career to be a mother? Are you a full-time carer? When did you decide to put your life to the bottom of the pile?

Why have you decided to dust off that dream and give it an airing now? Are your children grown? Do you consider it 'your turn' now that everyone else in your life is doing what they want?

Whatever your reasons, it's never too late

When, at the age of 28, I told my family I was retraining to be a plumber, they said it would be too hard, too heavy and too difficult for me to get work.

My friends said, 'wow!' 'Oh my goodness' and 'You go girl.' (Also, 'Great! I need a bathroom fitting') But I had already decided that's what I wanted and decided to listen to my pals' cheers of encouragement rather than my family's words of caution. Yes, it was scary, yes it was daunting and yes, it was financially a risk. But, yes it was exciting, yes it was an adventure, yes it was emotionally rewarding, yes it was a success, and yes I am now in charge of my own destiny. It didn't take long for me to reap those rewards. Within a year my diary was full and I had a waiting list three months' long. These days I cherry-pick the plumbing I do myself and have turned my attention to changing the world for women in the trades.

If you are reading this book because you want to reinvent yourself right now you may be worried that you are too old to be a plumber. Most women reinventing themselves as plumbers are between 25 and 35, with a lesser but still important percentage from 36 to 45.

Society would have you think that you couldn't do the job in your thirties, but I and the many women I interviewed when writing this book would dispute that strongly.* These are women who have left their 'comfort zone' and 'devil they knew' for a life more fulfilling. They are the role models for the future just as YOU will be.

Since you haven't decided that reinvention is not for you and you are still on this journey, let's take a look at reinvention itself.

The best reinventions come out of a passion. We know that necessity is the mother of invention. The need for shelter means we had to invent a way of cutting and shaping the trees. To cover our heads in the rain we had to invent umbrellas. To make ourselves into the perfect wife we had to invent ourselves and reinvent ourselves as the perfect mother when we had our children. Then we reinvent ourselves again to get back into the workplace. We do all this without thinking. We do all this because it is our role to do it and we love to fulfill our role, and we want to be the best we can. We also understand that this will change. But our lives do not end once our children are grown, and if we don't have children our lives are not meaningless. The 'work in that job until you retire and then die' no longer applies. These days we are constantly being fed 'live the dream'. No matter how fulfilling it is to bring up your children it is time limited and it comes to an end. Your children grow up and leave home.

In some the need for reinvention is so strong it can create health problems, as in the case of my colleague. In many cases we are pushed to such extremes, we hit such a low that we reinvent ourselves to survive. In society we are encouraged to reinvent ourselves in certain ways and actively discouraged in others. When I visited some rural villages in India I saw how perfectly acceptable it is there for women to build roads and dig wells and those same strong women being afraid to speak out in public. Here, in the most unlikely of situations I witnessed reinvention on a massive scale. As a result of the empowerment work done by The Hunger Project** (A global charity that I had raised $50,000 for and was visiting to see some of their work) illiterate women some of whom had never left their villages, blossomed into leaders, bringing schools, water, roads and electricity to places that had never had it before. To be a witness to this was a humbling and inspiring experience. Reinvention is certainly not restricted to the West.

Some reinventions are simply a result of unexpected happenings. Karen, a very successful plumber was happy in her job, until she needed a new bathroom and decided to 'have a go' at fitting it herself. She loved this so much she decided to train and is now controlling her own life as a plumber in her own business.

I never dreamt of being a plumber. I dreamt of owning my own life, of being in control of my own destiny. I couldn't have imagined how becoming a plumber would change my life at the time. I couldn't have imagined how fulfilling that reinvention would be, and that it would be so profound that I am now dedicating my time, resources and energy to bring that to others.

Ask yourself what you want your life to represent. It may be a 21st Century notion to follow your dream but only because for the first time in 200 years have we been able to say that we are no longer happy to work in the factory, that we want to own our lives.

I stayed as a teacher for another whole year while I figured out how my reinvention would manifest. I battled with the clichés of 'comfort zone' and 'devil I knew' With the uncertainties of jumping, and the crises of my pal confidence. But in the end, this jump into the unknown was coming and I had to do it. My fulfillment was worth more to me than the salary, the car or the 'trappings', and it has given me opportunities that never would have come my way if I'd stayed as a teacher.

With the benefit of 20/20 hindsight I realise that what I was doing during that year was completely turning my life upside down inside out and round the bend. I had reached a stage in my life (and my relationship) that meant something had to change. I can't really say whether the change in my career caused the changes in my home life or vice versa, but until then I was a like a caterpillar turning into a chrysalis. I felt drawn to this change, just like a caterpillar and just as a caterpillar has no idea what it will look like, I had no idea what my future would look like. My emergence from the cocoon was to take a little longer than that of a caterpillar but the beauty that resulted was greater than I ever could have imagined.

You have probably guessed that I am a huge fan of reinvention. I am also a huge fan of living the life you choose. By this time, I imagine that you know I would love you to choose plumbing to be the vehicle that takes you to that life. Largely because I will get to meet you should you choose plumbing, but mostly because women have fought for years to get to the point when they can say 'this is what I want for my life'

Find your dream

Pick your dream, and find a way to live it. For me, being a plumber has meant I can travel the world, meet some of the greatest entrepreneurs of our time, speak to women all over the world and encourage the next generation of female plumbers, help the cause to bring fresh water and sanitation to the world, leave the legacy I want to leave and do some pretty awesome work with pipes and my beloved water.

You may be in a position where your own wants and needs have been buried so deep that you don't even know where to begin on the reinvention process.

This was the process I used and it may help you to get to your reinvention.

If you could have one thing/feel one way and die happy what would it be?

For me it was peace of mind. That is my one thing, but what do I mean? What constitutes peace of mind for me? I break it down like this;

- *To know that I had been the best person I could be*
- *To know that I had lived a good life and been good to others*
- *To know that I had no regrets, that I had taken every opportunity I could have*
- *That I had caused no harm to others*
- *To know that I had lived a life that I wanted to live, with the people I wanted to be with*
- *To know that I had inspired others*
- *To know that I was loved and that I gave love.*

That was and still is how I define my life and my life's work. I can call this book The Joy of Plumbing because being a plumber has and is helping me to live to those high standards.

To find these things out about yourself, put yourself in a place where you can be truly alone with your thoughts. It's OK to remember your childhood dreams and it's OK to have moved on from those.

Remember, you are alone and nothing you think or feel is disallowed. Guilt has no place here. If you have regrets now is the time to think of those and let them go. Accept that that was then, and this is now. You can get your very own guided meditation to help you to get your reinvention plan started*.

If guided meditations are not your thing and you just want to be able to use a tool that will help, then I recommend the fantastic system of Wealth Dynamics. Wealth Dynamics helped me in my most recent reinvention from plumber to plumbing and business mentor. The system is based on ancient knowledge but says that each one of us has a path of least resistance to discovering the wealth inside us. The test is devised for individuals who feel that they are in the wrong job, doing the wrong thing, or just plain feel that they are not reaching their potential. I discovered Wealth Dynamics in 2006 when I was trying to find a way to stop working in business and work more on it. The 15-minute online test showed me where my strengths lay. Things that are obvious to others observing me but not so obvious to myself. Knowing my Wealth Dynamics profile gave me the map I needed to plot my onward journey. I loved the system so much that I took the opportunity to study it in great detail with the creator of the system in Bali. I now use the system to inform business decisions and help in my mentoring***.

So you are on your way to reinvention. This is an exciting time.

If this book has inspired you to continue on with reinventing yourself as an ace woman plumber then I am very much looking forward to meeting you one day. If it has inspired you to reinvent yourself anyway but not as a plumber, then I am proud. Although I chose plumbing as my vehicle for getting the life I love, I fully accept that it may not be the way you choose to reinvent yourself. The key is to continue the journey you have started, to find ways to keep inspiring you to move forward and to revel in your successes.

I believe we are ready to embark on the next leg of the journey. Having started the reinvention process you will probably want to know how I can be so confident in telling you that your business will succeed. I have this confidence because I have seen it over and over again. As much as 70% of women reinventing themselves as plumbers will do it in their 30s, and stories from the women plumbers I have spoken to and met from the UK, USA, Canada and Australia have all borne this out.

In the next leg I shall be looking at the differences between having a job and being in business for yourself as your own boss.

Web links

To join our online community visit **www.stopcocks.co.uk**

*You can get a download of the interviews and your very own guided meditation by visiting **www.joyofplumbing.co.uk/downloads**

** The Hunger Project **www.thp.org**- Empowering women and men to end their own hunger

***Get your wealth dynamics profile here **http://www.8pathstowealth.com/Profiles.html**

CHAPTER TWO

Starting a Business Versus Having a Job

The Three Cs

Right up until the 1990s it was still generally believed that you left school, got a job, worked in the same job for 50 years then retired. Even though that was rapidly becoming not the case at all. So rigid was this belief that even in the mid to late 80s when unemployment in the UK was at its highest ever at 3m the result was a broken population. The divide between rich and poor widened ever greater as yuppie culture hit the UK. On one side stood the ultra-rich stock market and city traders and on the other stood the broken factory workers, miners and manufacturers.

Technology had created massive riches for some but put the majority of others out of a job.

Just about this time I was deciding to leave my job and start my own business. 'Crazy!' you might say, but that just goes to show that when you put your mind to something, there's nothing that can stop you.

Take a moment to think about that.

How many times have people said to you

> **'That will never work'**
>
> **'You'll never be able to do it'**
>
> **'That is a terrible idea!'**

The list could go on, but you see what I'm getting at. We are surrounded by people who think we will fail, people who 'know better' and people who would like to see us fail so they can say **'I told you so.'**

That is what I was told when I broke the news to my family. Oh no, not only had I gone against my tradition and rejected marriage for two university degrees and a career, now I was dumping that career and doing of all things plumbing! I can sympathise but despite all the odds that was what I wanted to do.

Let's look back at the three Cs; **Confidence, Competence, Character**.

I definitely showed Character in making that decision. Was I confident? No. Was I competent? Well, at plumbing per se, no. I thrived at college. It took me back to when I was at school wanting to study metalwork and engineering. I quickly became Competent at handling the tools, joining metal, constructing systems that held water. I believed that in order to build my confidence I needed to be in a job. Imagine my dismay when firm after firm turned me down even though I passed my exams with distinctions. My Confidence was at an all-time low until I needed to do some work in my own house. I realized then that I didn't have to rely on an employer to pay me, that I could earn my own money, be my own boss, decide my own destiny. It was like a cover was lifted from my eyes. I still had to work on building my Confidence, but that came just because of the different situations I was meeting every day.

Now I had the full package; Confidence, Competence and Character. That was me. But what about you?

Lets look at the own business versus job question in more detail.

What about the security of a job?

Whereas it may be true that in a job you know where you are. You get up every morning, do the same routine, shower eat and out. Work all day then home eat and bed. You may get some weeks off when you can go on holiday to spend your hard earned money. Is this you?

It's true that you have a steady income, a monthly salary that comes in month-on-month and that, if you look after it correctly, you could actually save up for 'a rainy day'. If you are in a job right now take a moment to look at your life.

Are you doing your job because you love it or because you have to pay bills?

Does your job make you want to jump out of bed every morning with anticipation of the day?

Do you love the company of your colleagues and can't wait to see them everyday?

Does your job require you to stretch your mind?

Is your job character building?

If you have answered yes to more than one of the questions above then I want to be the first to congratulate you. You have a fulfilling job and that is great. I want to say thank you for reading this far and you probably don't want to read any further. Your journey ends here and I hope you've enjoyed what you've read. Hey, you're still welcome to come along, just in case there's something else in life for you. Let me give you a minute.

You're still with me, so you probably want to know what's next.

Why did I leave a perfectly good job with very promising prospects, a good salary and job security? (Teachers in inner-city schools are assured of long contracts) for the life of self employment, marketing my own services, having dirty hands and fingernails, and building a business from scratch?

Simply put. I stopped loving the job. I stopped wanting to go into the same place day-in day-out, seeing the same people, having the same routine. Even though I consider myself as someone who needs security, I was willing to give up 'security' for 'uncertainty'. My regular salary and 'safe' job met my financial needs, and my need for Confidence and Competence, but my Character needed more.

I can represent my level of satisfaction with my work life in a pie. I call it my circle of fulfillment. This is what my circle of fulfillment looked like when I was teaching.

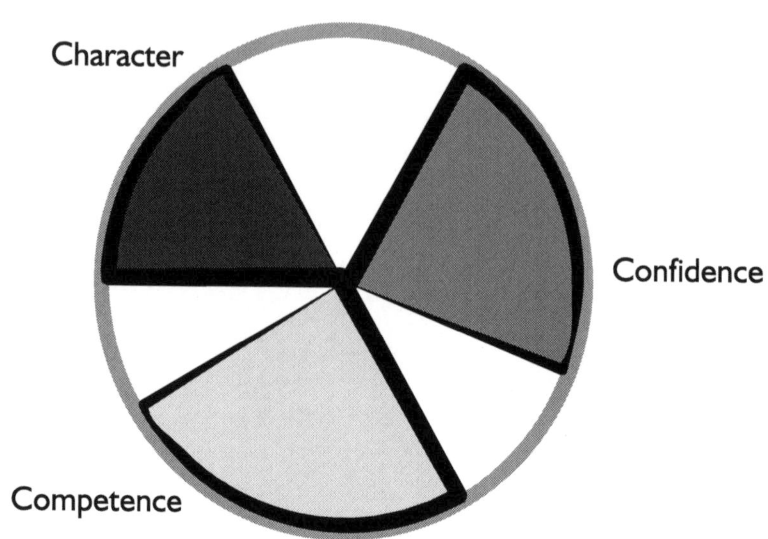

Far from full, but at least my pieces of pie for Competence and Confidence were over half filled.

Take a look at your circle of fulfillment for your working life at the moment. In fact I've given you one that you can fill in for yourself below. Be honest now.

If you don't like to write in books, (and by the way, I am one of those) you can go to the website and download a whole heap of stuff including a fulfillment circle sheet*. Before any journey it always helps to know where you're starting from and I found this a really helpful exercise in 'getting real' with my situation.

If you don't mind writing in books then go ahead and fill in the circle for yourself.

You can use the following questions to help you.

Confidence

How do you feel you rate your levels of confidence in your everyday work life?

Do you feel you can handle most things that come your way?

What does it feel like to be the last person to arrive, even if you're on time?

Do your colleagues come to you if they need help with any aspect of the job?

Do you see what needs doing, and do it?

When you answer the phone are you apprehensive about what the call may be about?

Do you put yourself forward for tasks such as the office party, or collections for a colleague's leaving present?

Do you speak up in meetings, or are you more likely to just listen?

If a 'controversial' decision has been made, do you voice your opinion or do you go with it quietly?

Competence

How do you rate your levels of competence at your job?

Can you handle most things that arise at work?

Are you the 'go to' person when there's a new task to perform?

Can you adapt your skills quickly to suit the task at hand?

Do you solve problems quickly and with minimal fuss?

Have you ever been 'Employee of the Month'?

Do your colleagues recognize you as brilliant in your job?

Are you the person people come to to discuss aspects of the job?

Character

How does your work life build up your character?

> Do you feel that you are properly appreciated for the work you do?
>
> Do you feel that you are paid adequately?
>
> If there was a crisis at work, would you be willing to put yourself out and work extra hours?
>
> Do you feel like you are a 'different' person outside of work?
>
> Do you consider your work colleagues to be your friends?
>
> Are you working to live, or living to work?
>
> Do you 'make a living' or 'have a life'?

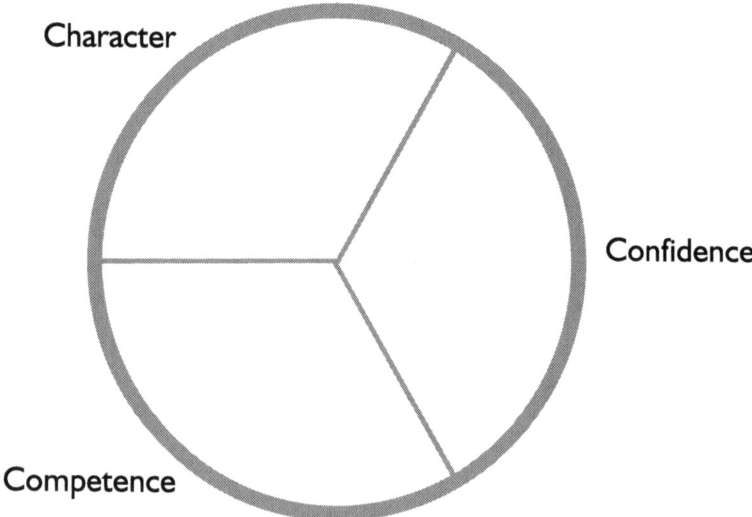

Now you know where you are. You have taken a good hard look at your feelings towards your work life and you have decided to stay on the journey. I'm glad, and I hope you will create a fabulous journey for yourself throughout this book.

Let's have a look at the issue of confidence in more detail. This has always been a stumbling block for women trying to break into a male dominated profession such as plumbing.

Some of the questions I am asked on a regular basis is

Will people hire me?

Will I be able to do the job?

Will I be successful?

These questions are so fundamental to starting your own business that I have dedicated an entire chapter to it called 'Why do Customers Love Female Plumbers?'

The deeper issue here though is that of confidence. Take another look at your circle of fulfillment. If you have a mostly filled in section for Confidence (over 70%), then you absolutely need not worry about the questions above. No, honestly. I can promise you that your business WILL succeed.

How can I say that with such confidence? Because a confident person, knows that they can do whatever it takes to succeed. If you have 50% or below, then I can still tell you that your business is likely to succeed, and that you need to build your confidence levels up. There is an ancient Chinese saying:

'To know and not to do is not yet to know'

If your Confidence score was 50%-ish then simply biting the bullet and going for it will increase your confidence levels.

One of the biggest issues around women going into male dominated areas like plumbing or electricianing (to coin a term) and something that I receive most concerns about is that you won't have the knowledge required to finish a job. If that is you, then please stop worrying about it right now!

I was speaking to a female roofer friend not long ago and she said that she would have gone into plumbing but there were too many things to keep in her head at once. She was referring to the hundreds of different products, system designs, regulations and all the different ways to put all those different things together.

She had made the same mistake I made when I started and the same mistake I am hoping you won't make.

You don't have to know everything before you can begin

I want to emphasise that. Yes there are hundreds (if not thousands) of products; taps, bathroom suites, boilers, fires, radiator valves. There are hundreds of different types of all the above. These are just designs. Most often, the differences are simply cosmetic. The working principles have hardly changed in the last 100 years. Take the inside of a WC cistern for instance. That technology has hardly changed since it was first invented in the 1800s. Some innovations have happened, sure but they rarely affect the fitting.

Many new plumbers that I meet who have decided that they want to work for themselves say, 'I've never fitted one like that before'. Now, more than ever, bathroom, kitchen and heating design are in the spotlight. It is more and more likely that you will come across things you haven't seen before. The swap from two taps on sinks to the monobloc took very little time to assimilate itself into design.

There are some principles which, once learned, can be applied across the board.

I will simply list some of those here as they apply to plumbing but certainly the same rules apply to any of the other trades too.

Plumbing is one of the most diverse of all the trades. It is beyond the scope of this book to go through the principles in much detail. Those details can be found in a series of tutorial videos that I have created**.

Principles

1. *Joining pipes*
2. *Different hot water systems (combi, cylinder etc)*
3. *Different heating systems (sealed, gravity, pumped)*
4. *How different types of systems work (direct, indirect etc)*
5. *How different types of valve work (screw down, quarter turn etc)*

6. *How water is provided to the property (mains, spring)*
7. *How to identify different pipes (heating, hot/cold, mains etc)*

Once you have grasped these principles and you have the Confidence and Competence to apply them, you will be able to tackle 90% of the things you are presented with.

Of course it takes a confident person to say, 'I'm not sure about that, I'll have to look it up'. Far from being seen as a failure, customers love honesty and thoroughness. They would much rather you took advice and got the job right than muddle through with the possibility that it might go wrong and end up costing you and the customer dearly.

Here is where having someone you can talk to is invaluable. You may have guessed that I love to talk about plumbing, and I'm not the only one. Back in 1990 when I trained there was George. He had been a time-served plumber for 50 years and was working as a technician at the Leeds College of Building. At first I felt nervous about asking questions but when I realized how much he loved talking about his beloved trade and how he enjoyed our conversations as much as I did there was no stopping me. Just knowing I had George on the end of the phone (there were no mobiles in those days and he very kindly gave me his home number) meant that I could take on the most challenging and demanding of jobs. Most notably was a 15 school radiator heating system which was to be fueled by a wood-burning stove in the basement kitchen, and incorporate an interchangeable gravity and pumped hot water system with the cylinder being fed by 28mm pipework.

WHAT? I hear you say…….. Exactly! That's how I felt until I was able to break it down to the basic principles.

This is how the system works, this is how to join pipes together. George was so excited he even came to the property to check it out. I installed the system almost totally single handed, (only towards the end and much to my relief, a Parisian plumber friend called Claude, who was on holiday, but couldn't resist), in the freezing cold winter of 1991 bending 28mm pipes by heating them up and jumping off a bench to get enough weight onto the bending machine.

If Confidence is your issue, then think back to other areas of your life. I'm sure the first time you sat behind the steering wheel, you were not confident at all. Now, you hare up and down the motorways like you've been doing it all your life. Remember:

To know and not to do, is not yet to know

Learn the principles, get a mentor, and practice, practice, practice.

Decide your destiny, intend your legacy

We all leave a legacy. Even if you don't wish to leave one, you will. When I started my training in London, (the only female, of course) my fellow students were there on the whole as a result of the media hype and the £60k plumber rumour. When I moved to Leeds all the plumbers on my course were the '& sons' of their family business. (Incidentally, just four mies from where I live, there's also an '& daughter' plumbing firm) Legacy has always been associated with business. Trends towards creating a legacy have changed. It's true that most of us want to do the exact opposite of what our parents did. (My father was an electrician and that definitely decided me against that) In my culture (Turkish Cypriot) it was generally believed that I would marry as soon as I was legally old enough and continue the family line with children and husband. It wasn't even on the cards that I should continue an education. (As indicated by my mother when she stated one day 'I don't know why you're bothering with school, you're only going to give it all up when you marry'. You can imagine my face)

Happily they are over the disappointment and have moved into proud and I can continue to create the legacy I wish to leave.

At the risk of being boring and repeating myself; I feel that I cannot hold the knowledge inside. I cannot stand idly by and watch while people die from unnecessary water-borne diseases and I cannot grow old comfortably without doing what I can to make a difference in the world. To that end, I see my life's work from now on to be set towards creating a nationwide community of female plumbers, and that Stopcocks becomes a household name for providing quality and trustworthiness through the network of self-employed female plumbers that will be generated by The Stopcocks Business Scheme. The growth of this formidable group of female plumbers will enable me to turn my attention to global water issues. After all 'water is life' and you know how I feel about water.

So that's what my legacy will be? I find it comforting to know that I am not simply living out my existence. That I am purposeful and on purpose. This is how plumbing has freed my life. I do not go from day to day longing for retirement, when I can do all the things I want to do, but be too old to do them. I want to live my life now and when I think of the future it fills me with joy. This is The Joy of Plumbing. So you can see, it really isn't about joining pipes. It's about the Confidence, Competence, and Character that plumbing has afforded me. It has allowed me to fully explore all the aspects about myself that I love and that I will help you to explore in yourself in later chapters. It's about having the Confidence to take on a challenge, the Competence to carry it through and the Character to see beyond.

What will your legacy be?

It's not a difficult question if you have children. Most parents would say they want to make sure they can leave their children with something when they die. You will definitely leave them something when you die, even if it is only a tiny glass ornament that was left for my friend when her mother died. (She has it on her mantlepiece).

Don't make the mistake of thinking it's too late to create something special. I was 28 years old when I started plumbing and many of the women on the scheme are in their thirties. It is never too late. I'm not suggesting for a moment that you force your children to be plumbers. When your life is free you can turn your attention to whatever you like. Look at Bill Gates. He's decided to give a significant proportion of his money away. For him, and for me (you too can have something in common with Bill Gates), our most precious asset is our time, and we don't want to waste too much of it doing anything we hate. We all have to do things that need doing, and to make the pill less bitter I use a technique that I will describe later on. The point is to make everything you do something you enjoy doing because it will give you the freedom to do the other things you love, before you are too old to do them.

I learned to scuba dive when I was 44 and windsurf when I was 46. I still hope to learn to snowboard and ski.

Why is now the best time to start a business?

Duncan Bannatyne, best selling author, multimillionaire, star of Dragon's Den and entrepreneur, said 'Every minute is the best time to start a business'. I agree. (But then I would) He started his empire from a phone box.

As I write this book the UK is just emerging from one of the worst global recessions in history. Very scary times it seems. Yet this is the precise time I chose to step up my game. Many small businesses have suffered in this terrifying economic climate but those that stepped up when everyone else battened down are the ones that are thriving now. Those that hid to wait the storm out have found that their 'loyal' customers were not so loyal in the end. Larger orders have been shrunk and have gone to the players still in the market. Those that prevailed are the ones that offer the best customer service and the highest quality of work. This is the reason Stopcocks didn't 'feel' the recession. People will always need plumbers. There is and there will only ever be one way to move water and that is through pipes. Recession or no, when a plumbing job needs doing, it needs doing. It's no good telling your pipes to wait until the recession is over. They have been working hard for years and when they need attention, they are worse than the most demanding toddler.

Let's look back to 25 years ago.

Business was stereotypically the domain of stuffy men who owned factories and treated their employees less than kindly. A 'them and us' culture existed and Britain was reeling from Thatcherism and the breaking down of the trade unions. Entrepreneurs of note were Richard Branson with his Virgin brand and Anita Roddick with The Body Shop.

All the while in Silicon Valley, California, technology was being developed that would change the face of business and the world in ways we could hardly imagine.

I first got a computer in 1987. The internet was still a dream.

To start a business you needed a tome of a business plan. You had to research your market, your competitors and your geography, by painstaking library and footwork. Then you went to the bank to ask for capital and they would put you through some more hoops of fire, and tell you all the reasons why your business would fail, before refusing you the loan. So then what? You went to friends and family for startup money, who would find it hard to refuse you. Is it any wonder most businesses fail within the first year?

Why would they fail? They were either not researched well enough and failed because the market was not there or they failed because they were not thought through enough and couldn't keep up with demand. Perhaps they didn't have enough budget to keep jumping through the hoops of marketing, sales and follow through.

So what's so different now?

One word answers that question.

Technology

The world-wide web was only used by 2.1m people globally in 1990. That figure had increased to more than 45m by 2009

In 2006 only 50% of the UK population had internet access. This leapt up to 70% in 2009 2m households more than in 2006.

It has never been easier to research than it is now. Never easier to be competitive and never easier to get the word out to your market.

In 2008 I decided that I wanted my legacy to be that women could be successful plumbers all over the UK long after I had put my spanners down. I set about researching the best ways to achieve it. I quickly found that in some institutions the numbers of women were so low the college didn't even bother to document it. The number of female plumbers in the UK had grown since I started but still remained pitifully low. Yes some laws had changed, and more women were entering colleges. (I knew this because of the increasing number of emails in my inbox from trainees) but still the number of actual plumbers remained static.

Even in 2008 technology in the UK hadn't caught up with what I was trying to achieve. By using the internet I was able to locate the plumbers, and was even able to pass work to them as it came up, but this still didn't satisfy me. It was the coming of the social media revolution that really made it possible for me to begin putting my dreams into action.

All of a sudden it's possible to get reactions straight away, no need to wait for days or even hours. One particular case comes to mind where Twitter really got a young woman's plumbing career off to a good start. She joined the business scheme and within a few days the Stopcocks machine set into motion. Just by using the social media forums of Twitter and Facebook we were able to not only get her noticed by potential customers but also by the local press. Within the first two hours of posting she had already received three requests for work and set up an interview with the local press, giving her business the boost it needed.

Suddenly it has become so easy to set up in business and get to your customers immediately. The impact this has on starting a business is immense. Right now you can start your business and begin to attract customers with minimal financial investment.

Things have moved a long way since I started in 1990. Letting customers know you are there has never been so easy. To achieve my dream of having a Stopcocks plumber in every major city and town in the UK by 2015, meant employing all these techniques to the fullest extent.

How can I be so confident that your plumbing business will be a success?

It's no surprise when you take a look at what women have been achieving throughout history.

There are countless books and studies that document this and can be easily found in libraries bookshops and online.

In the next leg of our journey I'll be asking 'Why on Earth do you think your business won't succeed?' and I'll be sharing some stories from successful female plumbers from around the world, and asking the big question:

> **Why do women have such confidence issues and what's it really all about?**

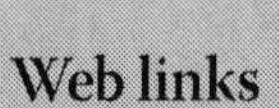

*Blank fulfillment circle sheet and interview download from
www.joyofplumbing.co.uk

**Tutorial videos at
www.youtube.com/stopcocks

CHAPTER THREE

Get out of the way

Whatever makes you believe you won't make a brilliant plumber and that your business won't be a success?

Let me ask you this instead. How come whenever I am in a public setting – it could be a party, a conference, a pub, or in a supermarket queue – I always end up giving my phone number to someone who needs a plumber?

We are always in demand. There is only one way to move water and that is through pipes. As long as that is true, then we will always be in work. I've dedicated a whole chapter to why customers love women plumbers so I won't go into it here. Suffice to say, that it doesn't matter where you go, when people know you're a plumber, they want to tell you a story about some plumbing incident or some bad experience, and then they ask for contact details.

Some time ago I was having dinner out with some friends and I overheard the family on the next table talking about a problem they were having with the plumbing in the new house they were trying to move into. Although I tried not to eavesdrop I couldn't help but prick up my ears at the various problems I caught snatches of. In the end I leant over and said, 'It sounds like you need a plumber.' With relief they said, 'Yes, do you know one?' I handed them a card and arranged a visit. Now that achieved two things. It put their mind at ease, and meant they could carry on with their evening and it got me some work with the potential to earning more with the recommendations they would provide to their friends. I went back to my meal.

We can come back to the old chestnut of confidence. I could hear what the problems were and I knew that I had the knowledge and skills to put it right. I was confident enough to speak up and it paid off. Later on I'll tell you how I made the decision to leave teaching and become a plumber.

Confidence

I'm not going to deny that I had confidence issues. I can hear you saying, 'it's alright for you, you're confident,' but I have to tell you that I had similar issues when I first became a plumber. I felt I had to prove myself to the guys out there and in college I strove for perfection. I had no role models then. Speaking to women since it seems that those things haven't changed. The tutors of female plumbers (or any of the nontraditional roles, for that matter) have said that it is the women that are most dedicated and they often put a lot of pressure on themselves to do so well that they put the guys on the course in the shade. After 20 years of studying this phenomenon I can confidently conclude that the reason you study so hard and want to do so well in college is directly related to levels of confidence. I would go even further and say that we want to show ourselves and our tutors that we can do it, despite what society says we should be doing. I don't know about you, but I took my workshop samples home with me. My very first herringbone lead weld was a thing of beauty, that I was very proud of. Don't get me started on the fabulousness of bronze welding.

We will no doubt meet our old friend 'confidence issues' some more as we travel on this journey. When we do, we will stop and visit for a while, pick up some new bits of insight and go on our merry way.

No Accident

At the beginning of this book I said it was no accident that I would write it and no accident that you are reading it. What did I mean by that and why did I say it?

Let's put things in perspective here. Let's take a look at the big picture.

Most women won't want to become plumbers or joiners or electricians. In fact, most men won't want to either. Most men won't become nurses and neither will most women. Most people won't want to work in the trades, or in the caring industries. Most people, of either gender won't become or want to become politicians. We can say that none of the roles listed are traditional, because there is only a minority of people that will want to spend their lives doing them. Many people at some time in their lives have wanted to be a pop or film star. But most people don't end up as them either.

You are already in an elite group of individuals. Less than 1% of plumbers are women in the UK, and the stats are quite similar in other Western economies.

I want to dedicate this part of the journey to persuading you that you are doing the right thing and that your plumbing business will be a success. Well, when I say persuading you, what I really mean is giving you the benefit of my experience and that of the 50 or so women I have interviewed for this book, who have all without exception said that they don't regret for one moment making the decision to stop what they were doing and become plumbers working for themselves. We have all collectively talked about the hardships of going against the grain and the great victories we have felt as a result of that.

The Bigger Picture

I am one who likes to look at the bigger picture and in this case that means having a real hard look at why you might be thinking your business will fail.

I'm going to imagine that you are at a point where you have completed some course and you have been a regular attendee. I'm going to continue by imagining that your course is in the manual trades, plumbing in fact (although it could be any course at all, non-traditional, in any trade, or not) I am also going to assume that like almost all women reinventing themselves, you have not slacked off and that you have worked hard at college.

My personal opinion is that you will probably have performed at one of the highest levels amongst your peers, although I won't be adding this to the evidence. Think about your time in college and what your tutor had to say about you and your work. I passed my exams with distinctions and this is something I have in common with nearly all the plumbers I interviewed for this book.

Let's assume you have been to a college or a training centre where you have been handling tools. Your course has been anything from six weeks to three years long. You have learned about joining metals together by soldering and welding. You know how to prepare copper and use a blow torch. You know the different metals, used in plumbing and have even bent pipework using several different methods. You have probably done some stuff you will never use again like bending gun metal pipes using heat and sand, created a thread and joined the pipes together using steel fittings. You have done lead work and know how to make flashings for a chimney using lead, aluminium, and maybe even copper. If your course was like mine, you will have created a joint in copper by drilling a pipe and creating a shaped piece to fit the hole perfectly and joined them together using a particular method of bronze welding.

This is just a short list of the areas you will have covered and not to labour the point too much, you already know more than most people by this stage, and you haven't even started working yet.

The next assumption I will make is that you like to take pride in your work, that it matters to you what people think of you and that you would like people to be happy with your level of work. This not only applies to your plumbing work, it probably applies to any work that you do and any work you have done in your life right up to now. You care that your level of work is high. You worry that you have remembered every thing you need to do, and that no stone has been left unturned in your attention to detail.

'Hold on!' I hear you saying, 'What has this to do with my business failing?'

So far all I've done is talk about how brilliant you are at what you do.

Karen Matthews has been plumbing for three years at the time of publishing. Before that she was in advertising. She needed a new bathroom and decided to have a go at fitting it herself. She enjoyed it and decided to look into becoming a plumber.

She studied for three years part time, while continuing in her 9-5 job, and practiced her skills in her own home and those of her friends and relations at weekends and evenings.

To start her own business she cut down her regular hours and worked part time plumbing and in the office. After only six months she had a client base large enough to leave her job and 'plumb' full time. After only two years Karen became able to employ another worker to help with the load. To illustrate how successful Karen has become; we had to reschedule our interview four times and it took more than six weeks for Karen to find the time to give the interview.

Karen states that even though she is a successful plumber with a large customer base and a waiting list, she had the same doubts you might be having. She was nervous, and anxious that the stereotypical views would work against her. She wondered whether to advertise herself as a woman plumber or if that would turn the customers off.

She found (as did I) that customers were very happy to hire a female plumber and once they had hired her they would be most likely to call her time and again.

Rose, a new plumbing trainee, believed that she would go into an apprenticeship, and 'learn everything' before striking out on her own. The truth is, you will never know 'everything' and after all my years I still don't 'know everything'.

You no doubt have heard the saying from the famous sports shoe 'just do it'.

My Story

In 1990 after I had applied to every plumbing firm in Leeds to take me on, with my confidence low and self-belief knocked hard I was left with two choices. Strike out on my own, or go back to what I had done before and find a job as a teacher. This is what I did and perhaps this will help you too.

I took a piece of paper and I divided the page up as shown below

Teaching		Plumbing	
Pros	**Cons**	**Pros**	**Cons**
Paid long breaks	Rigid	**Being my own boss**	**Irregular pay**
Regular pay	Routine	**Meeting new people**	**Sexism**
Easy	No challenge	**Fixing things**	**Dirty**

I wrote down all the advantages and disadvantages of doing both jobs.

I wrote this list until I couldn't think of any more and it soon became clear to me that this wasn't going to help me decide. I could always come up with just as many sensible reasons why I should go back to teaching as I could for becoming a plumber. This was not going as well as I'd hoped. I would have to sleep on it and come back.

In fact I didn't come back to it for some few weeks. I had put the list down on a pile of things and forgotten about it until I was clearing the table one night to make room for a meal I was cooking for some friends. It fell onto the floor and my initial reaction was guilt because I knew I had been putting off this decision. Then I realized what I had actually been doing. Why I was so resistant to making the decision.

I quickly grabbed the page along with a fresh page and wrote two columns as follows

Go Back	Go Forward
Feeling a failure	**Freedom**
Having regrets	**Deciding my own destiny**
Never reaching my full potential	**Leading the life I want**

Go Back - represented finding a teaching job as I always thought that finding a teaching job would be a backward move. There was nothing about teaching that made me feel I would be moving forward in my life and this was a big decider.

Go Forward - represented starting my own business as a plumber as nothing about this path felt like a backward move because I hadn't done any of it yet.

I realized that my decision was not going to be based on 'WHAT' I did, but rather 'WHY' I was doing it.

It was this realisation that enabled me to take my life into my own hands. That was the day Stopcocks was conceived. From that day I felt I had a purpose and a direction. Now all I had to do was get out of the way and let it happen.

The only thing standing between me and my future was me

I had doubts about whether I was good enough to sell my services to customers. I had confidence issues, especially since I had been so demoralised by the stream of rejections I'd received from the local plumbing firms.

In 1990 printing was an expensive matter and computer artwork was a skilled job. My first leaflet was a hand written A5 (so I could get two on one photocopy) piece that basically said who I was, what I did and how I could be contacted. Mobile phones were very expensive so I bought a second-hand answering machine from Leeds market to make sure I didn't miss any calls.

I spent the next week, walking up and down posting leaflets and waiting for the calls to come in. The one thing I did do though was state 'woman plumber'. I drove to the nearest motorway service station where I had seen a machine that printed out business cards £3 for 100 cards. With just 'Stopcocks, woman plumber', my strap-line 'no mess and no messing' and my phone number I was set.

I had no business training, I knew nothing about record keeping, or book keeping, or pricing and my plumbing experience had so far been all workshop based, but for me the reasons were clear I would be in charge of my own destiny. I would be free. Within the first two weeks the calls came in and within the first six months I had enough money to advertise in the City Life magazine. They loved me in the offices and I ended up doing some repairs for them once or twice. I built my toolkit by buying second hand tools from Leeds market and invested in an A-Z, a diary and an invoice pad. I later added an address book just for customers.

I could say my entire business started with a few leaflets, a telephone answering machine, and some second hand tools.

It wasn't all plain sailing. It was hard work. Every day I would wake up feeling terrified and excited. I relished the future that was coming and was terrified about it coming too. With every phone call I worried that I might not be up to the task, but couldn't wait to expand my knowledge and experience. This was really pushed to the limit within the first six months when I was commissioned to fit the huge central heating system that was to run off a wood-burning stove that I referred to in a previous chapter. It was to have 13 cast iron radiators and a cylinder for hot water. The customer was a friend and I felt confident enough to tell him that this was the biggest project I had taken in my young plumbing life. (It's fair to say, I was terrified). He said he was confident that I was up to the task. This project stretched all of my design muscles and tested my knowledge of systems and how they worked. I needed help and contacted George, the time-served old-school plumber who was working as a technician in the college. I had called on his expertise a few times in the past and he hadn't failed me. He was clearly delighted that I was calling on him again. He loved plumbing and had been working as a plumber boy and man until he came off the tools at the age of 62. He couldn't stay away so he got a job at the college as a technician. He said he loved being around the whole plumbing 'scene' and we had many hours of George sharing his experiences and me soaking it all up.

With this expert help I was able to complete the job with just a bit of help and some more confidence boosting from Claude, the Parisian plumber, who confirmed what I'd done; and a four-storey Victorian house that had been freezing cold for over a hundred years was transformed into a cellar kitchen with a wood burner and toasty bedrooms, with free hot water to boot. My confidence was through the roof.

I could have walked away from that opportunity too scared to take it on. I can honestly say that taking that job and completing it was a turning point.

I was called back to the house once when the wood burner had heated the water to such an extent that the solder melted. Parts of it had to be re designed and re piped but the corner had been turned and it wasn't scary any more.

In my fourth year I decided to add gas to my skills, which meant going back to college again. This brought new concerns and for a while every time I hung a boiler I would lose sleep worrying it would fall off the wall!

Of course the boilers never did fall off the wall. The cylinders never did implode. The central heating systems always worked, the radiators were always firmly fixed, the bathrooms always looked lovely. I was always able to cope with anything that leaked or came apart. All I had to do was get out of the way and let it happen.

The Big Question

Confidence is a huge barrier, and it often stops us in our tracks, but what is it really?

The big question is;

Why do women have such confidence issues?

I received an email from a young woman who had more than two years' experience working out in the field. She had carried out a number of jobs on her own, including a full house power flush, negotiated with site project managers and completed many bathrooms and wet room installations. Yet she STILL felt she didn't have enough confidence to go it alone. So what is it that stops us from just doing it?

Without fail, every plumber I spoke to during the writing of this book has said that their main issue just before starting was confidence. Yet every plumber without fail has said that once they decided to just do it, their confidence began to grow. This really prompted me to look into the whole confidence thing and I wondered whether it was confidence, expectation or a bit of both.

There is a well known and highly documented phenomenon called 'The Pygmalion Effect' This describes how expectations colour the way we think and act. This has been studied since the 1960s where children were observed at school. Those children of which there were high expectations, excelled, where as where there was a low expectation by the teachers the children failed to make the grade. Once this was noticed and teachers were able to modify their behaviour towards the children, grades got higher. But latterly these same studies have been carried out with adults in the work place. Where output from employees has been shown to be directly related to the expectation of employers. A praised staff member will feel more valued and therefore will work harder. One that is degraded will not be motivated to be their best.

This may seem a side issue but it fits in very well with minority dynamics. That is, us. We are a minority in the field of plumbing, so (our perception of) society's expectation of us is that we won't be any good. So, in order for us to overcome those underlying and hidden expectations we feel we need to over achieve and be super confident before we even step out into the male dominated world of plumbing. Pair that with the most old fashioned worlds of all (construction) and we have a mountain to climb. I imagine men going into a female-dominated field will have a similar experience.

It is always good to remember that you have already started climbing that mountain. It's not really a mountain, we perceive it as that. I perceived it as that when I started and every female plumbing student I have met perceives the same. You just have to remember that you are not climbing Everest in your slippers. You are standing in the foothills of the rest of your life. The more you do the more you progress. The higher you climb, the better the view.

Remember when you first arrived at that new holiday destination? You'd never been there before. How did you organise that trip to be? Did you do the same as you always do? Maybe book a hotel with all meals included. A pool for you and the kids and organised trips?

Or maybe you just looked at a tour guide and bought flights with the intention to sort everything else out when you got there.

The 'safe' first option is where you have it all worked out so the destination is neither here nor there.

The second 'adventurous' option, means there are unknowns; what if you can't find a hotel? What if there's nobody there who speaks your language? What if you don't like it? What if it rains?

For either option you would have made certain preparations. You probably looked up the weather probabilities and packed accordingly. You probably researched and obtained any permissions required. You probably packed enough clothes to last you for the amount of time you'd be away.

The point is; no matter which option you choose, there are things that you definitely knew BEFORE and were prepared for. This preparation meant that you could go on holiday confident that you would enjoy it even though it was something you hadn't done before.

Now let's apply this to our scenario of starting your own business, or as I prefer, taking your life into your own hands and controlling your destiny.

It certainly smacks of holiday option 2. How are you going to view it and how is this view going to affect your confidence levels?

Will you view it from all the unknowns and get terrified? This way unconfidence lies. If we look at anything from the point of view that we can't possibly succeed, then that eats away at our confidence and before we know it we're reaching for the package holiday.

If you enjoy the adventure, get a kick out of the unknowns, and can't wait till your next fascinating experience, then your confidence will grow in leaps and bounds and before you know it you'll be flying high, in control and free!

On this part of our journey we have explored what may really be going on when we say we don't have the confidence. I've shared with you the processes I went through to making the final decision to really grab my life by the horns and make it what I wanted it to be.

At this point I want to say that I am no more special than you. I had no privileges, no special treatment, no help and no encouragement to do what I did. I hope your situation is better than mine was. The probabilities of your success are far greater than mine were.

So I ask again:

Whatever makes you believe you won't make a brilliant plumber and that your business won't be a success?

Our explorations through expectation and confidence have clearly shown that the journey is hard, but that we have some responsibility, and if we just got out of the way, we can let it happen. We've looked at how we can bring our other skills to bear and shown that the rewards for taking those steps are so much more than just a plumbing business. They represent taking control of your own life, choosing your destiny and perhaps the most thrilling thing of all; freedom.

CHAPTER FOUR

You are today's news

A brief history of women in the trades

Although women have always been active in every part of society, massive change came about across the western world with the onset of war and this is where my potted history will begin.

As the menfolk left to fight it was the women that kept the home fires burning and facilitated the war effort by building, maintaining and testing the weapons of war. They not only worked in the munitions factories but also worked in the mills making the uniforms as well as in the fields growing the food that kept the country fed during those hard years.

No surprise then that after the war when the men returned women were reluctant go back into the home. For years the country and the war effort had relied heavily on the women back home and in the field. Though hard work it brought with it a sense of confidence and self-reliance that was previously the domain of the 'man of the house'. Women realised they were much more capable than they had ever been allowed to believe they were before and they liked it.

A brief history

Here told very eloquently and with kind permission by Professor Joanna Bourke, Professor of History, Department of History, Classics, and Archaeology, Birkbeck College, London

Beyond domestic service

Did World War One actually improve women's lives in Britain? At the time, many people believed that the war had helped advance women politically and economically. Thus, Mrs Millicent Fawcett, leading feminist, founder of Newnham College Cambridge and president of the National Union of Women's Suffrage Societies from 1897 to 1918, said in 1918: 'The war revolutionised the industrial position of women - it found them serfs and left them free.' The war did offer women increased opportunities in the paid labour market. Between 1914 and 1918, an estimated two million women replaced men in employment, resulting in an increase in the proportion of women in total employment from 24 per cent in July 1914 to 37 per cent by November 1918.

The war bestowed two valuable legacies on women. First, it opened up a wider range of occupations to female workers and hastened the collapse of traditional women's employment, particularly domestic service. From the 19th century to 1911, between 11 and 13 per cent of the female population in England and Wales were domestic servants. By 1931, the percentage had dropped to under eight per cent. For the middle classes, the decline of domestic servants was facilitated by the rise of domestic appliances, such as cookers, electric irons and vacuum cleaners. The popularity of 'labour-saving devices' does not, however, explain the dramatic drop in the servant population. Middle-class women continued to clamour for servants, but working women who might previously have been enticed into service were being drawn away by alternative employment opening up to satisfy the demands of war. Thus, nearly half of the first recruits to the London General Omnibus Company in 1916 were former domestic servants. Clerical work was another draw card. The number of women in the Civil Service increased from 33,000 in 1911 to 102,000 by 1921. The advantages of these alternative employments over domestic service were obvious: wages were higher, conditions better, and independence enhanced.

Post war

Anxiety for their menfolk in war, the pressures of employment, combined with the need to perform housework in straitened circumstances and the inadequacy of social services exacted a heavy toll. It also made the withdrawal of women back into their homes after the war less surprising. This return to full-time domesticity was not, however, wholly voluntary.

In many instances, contracts of employment during World War One had been based on collective agreements between trade unions and employers, which decreed that women would only be employed 'for the duration of the war'. Employed mothers were stung by the closure of day nurseries that had been vastly extended during the war. Reinforcing these pressures were the recriminatory voices of returning servicemen. As unemployment levels soared immediately after the war, anger towards women 'taking' jobs from men exploded.

Women were also divided, with single and widowed women claiming a prior right to employment over married women. For instance, Isobel M Pazzey of Woolwich reflected a widely-held view when she wrote to the Daily Herald in October 1919 declaring that 'No decent man would allow his wife to work, and no decent woman would do it if she knew the harm she was doing to the widows and single girls who are looking for work.' She directed: 'Put the married women out, send them home to clean their houses and look after the man they married and give a mother's care to their children. Give the single women and widows the work.'

The cat was out of the bag.

During both World Wars women kept the country going and were called upon to do jobs that previously were preserved for men. After the war (WW2) the situation for women became hard as the nation strove to employ the returning soldiers. Women were pushed more harshly back into the home and many single women found work in service. It appeared that things were going down hill for women again. After some period when millions of young women were recruited back into service, it became unacceptable to those that had taken the opportunity the war had brought to break away from those shackles.

It's not unusual for women to take on building roles. As mentioned earlier, in many developing countries women take these roles easily.

The fact is though, that progress is slow. After 20 years of my plumbing career, it is still news to see a female plumber.

Do I see a future where there are as many women as men in plumbing?

The answer is no. I believe there will always be more male plumbers, I believe female plumbers will always be news and frankly, that is exactly how I prefer it to be.

Right now, we are in the unique position of being a novelty. This is what makes us special, this is why every time a new plumber joins the Stopcocks Business Scheme they can be sure that their local press will be interested in them and they can get a little leg up in their business.

The novelty of female plumbers hasn't worn off in more than 50 years. There is no other novelty that has lasted so long as women in the trades. Even though we are growing in number, the intake of females into the trades remains low. Less than 3% of tradespeople are women and less than 1% of plumbers are women. This is really great news for us. The benefits have been obvious to me in my business and from the interviews I've done for this book it seems, for those women plumbers too.

In 1990 I was contacted by a well known national directory of a certain colour. They wished to sell me advertising in their book.

I thought I'd give it a go so I filled in the form and sent it off. A few days later I received a phone call from the directory to confirm that I had filled in the forms correctly. I confirmed I had, but was told that I wouldn't be allowed to state 'woman plumber'.

When I enquired why I was unable to use my unique selling point in my advert they replied that they felt it would not be appropriate and that I might be attracting danger by saying I was a woman. I never got to the bottom of this reasoning. I assumed it was policy for directories but in hindsight I wonder if it was just the man that had opened my application form. I withdrew the word 'woman' from my ad, paid a fortune and got absolutely NO business from the directory at all. From that time I stayed away from them, until some years later I got a call from a Thompson Local rep.

Plumbers 359

Probably The Only Woman Plumber In The Region
- All Plumbing Work Undertaken.
- Full Central Heating Designed & Fitted.
- Boiler & Heaters Serviced & Repaired.
- Kitchens & Bathrooms Fitted.
- All Work Guaranteed.
- No Mess & No Messing.
- Complete Project Management Services Available.

CORGI 175946

STOPCOCKS
CALL HATTIE
MOBILE 07798 825091

A B C D E F G H

He had been lunch in the next town and spotted my card on the noticeboard. Always on duty, he checked the directory to see if I was listed. Of course I wasn't, so he called the number on the card and arranged to come and see me. I told him the story of the other directory and he reassured me they had no such policy. They allowed me to have the ad I wanted in the words I wanted. (They were more cost effective too) My business had moved onto the next level and I had to create a waiting list.

Being a woman plumber and shouting about it was the key that opened the doors to places I couldn't have imagined.

I was contacted by BBC Radio Leeds. Their version of Woman's Hour featured a full interview with me. Since then I've appeared in many newspapers, including The Independent, and numerous BBC Radio shows, including Radio 4's Woman's Hour.

My advice to you is always make sure you advertise yourself as a female, woman or lady plumber. They all work.

You see the thing is, when customers call you, they know they are getting a woman and it doesn't shock them to see you when they open the door. On one occasion I arrived at 1pm as arranged and I overheard the customer telling her husband in surprised tones, 'She said she'd be here at 1 and she's here on the dot'. For me it was always important to turn up on time to a customer's house. I wondered if I was unique in this but again, it seems all the women I spoke to said it was important to them too.

So, we will always be today's news and I applaud that. Here is a way that we can turn the prejudices of society to our advantage.

So what if plumbing is a man's world?

Plumbing may always be a man's world and as long as it is we will always succeed because we stand out. In business it's those that stand out that get ahead. Those that have something different to offer, be it a better service, more value for money, or the unexpected gender will be noticed and remembered. Couple this with the fact that we do deliver a great service, we care that we do the job right and customer satisfaction is highest on our priority list, and we have success guaranteed.

Away from the trades we can find the same tactic being used by male hairdressers. In my sister's hairdressing salon it's big news that she has a male hairdresser. The customers love him. One of my close friends is a male hairdresser and even though he relocated from Manchester to a remote town in the Pennines his client base remains strong with some of his customers traveling from London every six weeks just to have him cut their hair. No one can say that hairdressing is a man's world, yet male hairdressers are amongst the most successful.

I asked my hairdresser friend if he had any confidence issues when he started out, knowing that he was in a minority. He admitted that he did in the beginning. When I delved deeper he revealed that his issues were not with his ability to cut hair but with his ability to be organised enough to run a business.

At school I steered more towards the sciences than business studies.

I had no idea how to run a business when I started but I did know that I wanted my service to be second to none. I worried about my abilities to plumb, my abilities to get customers and my ability to run a business. I found that my unique selling point was that I am a woman. Second to that, I made sure I delivered a service few other plumbers were delivering at the time. That service was built on respect for the customer, high-quality work, a good aftercare service and peace of mind that I would always be on hand to sort out any problems that came up.

Instead of seeing our gender as a disadvantage, we can view this as the most powerful and useful tool in our toolbox.

Todays News

Web links

Read the articles in full here **www.stopcocks.co.uk**

CHAPTER FIVE

Why do customers love female plumbers?

Feeling Loved

Research findings

In a 2010 report entitled 'The most common and costly mistakes made by tradespeople' Andrew Priestley conducted research amongst customers of a British home improvement centre which also ran a centralized booking service for tradesmen. It asked, 'What are your key frustrations with tradespeople' The customers were asked to tell specific stories that related to specific tradespeople.

There was no prompting but the points to consider were these

- *the service being provided*
- *the critical incident or incidents*
- *the gender of the tradesperson*
- *perceived age group*
- *perceived level of expertise*
- *dissatisfaction rating on a scale of 0-5 ; 0 being very dissatisfied through to 5 being very satisfied.*

The results boiled down to 39 key issues. An extract of this report can be found at the end of this book (App I), for our purposes we shall only look at the first three.

Namely;

- *He didn't turn up*
- *He was lazy, didn't want to do anything more than he had to*
- *He didn't seem interested*

This was closely followed by, he didn't listen, he didn't know his stuff and he was impossible to contact. You'll notice the word 'he'. The report found some incidences where the issue was with a female operative, but these were very few and didn't make the top 39.

In my own research of 20 years talking to thousands of customers the stories seem the same. On one occasion I overheard a customer commenting with pleasure that I had phoned them back. Behaviour which, for me is standard, such as calling someone back, turning up on time, listening to the customer requirement, respecting the customers property, always following up with good customer care was sadly absent amongst my counterparts.

It seemed to me that customers had been 'trained' over the years to believe that they would be treated badly. That somehow, pride and honour in the trades had been exchanged for 'bodge it and scarper'

Bad boy syndrome

There were a couple of names that always came up in the staff room while I was teaching. The same people would have stories about the same children. Those children became known mostly for their bad behaviour. Could we have had a part to play in their mis-behaving? Earlier we saw how people adapt to what's expected of them. We expected the naughty children at my school to be naughty and so they were.

If we expect our tradesmen to be rude, demanding, bad at their jobs and 'rip off merchants' then do we not have some responsibility to bear? If they have trained us as customers to expect the worst, perhaps we could train them to be the best they can. I believe that everybody wants to feel good about themselves. I believe approval breeds approved of behaviour, and disapproval breeds disapproved of behaviour. This is what I term as bad boy syndrome.

How did this happen and what was I going to do about it?

I started to ask customers for their stories and I can tell you it didn't make for happy listening. The common threads though were similar to the findings of Priestley's report Add to that list; didn't finish the job and ran off with the deposit, left the place in a complete mess that took the customer weeks and money to clean up, did short days and spent most of it in the van outside and demanded constant cups of tea.

I was appalled to find that customers expected to be treated in this way by tradespeople.

At this point I have to say that I do not believe these behaviours apply to all tradesmen, and it should not be read as if I do. I have also heard stories of really great builders, electricians and joiners. In fact, I actively collected a group of other tradesmen around me, to support some of the plumbing jobs, and formed reciprocal arrangements, where we would call on each other when needed. Those honourable tradesmen have suffered discrimination as a result of the cowboy and rogue trader.

For me it was unthinkable to say I'd do something and then not do it. Unacceptable to agree a time and place and not turn up. It was common practice for me to call ahead to make sure the person was home before I turned up, and if I was held up, I would always call to let the customer know. Much to the customer's surprise I might add. For myself, and for the other women plumbers interviewed for this book, this is common courtesy and practice.

When I showed this report to several of the plumbers the look of shock on their faces said it all. Followed by gasps of 'I would never do that.'

Turning up is standard

The survival of any and every business depends on the customer. If the customer likes you and trusts you they will stay with you. They have gone through a selection process to pick the plumber for the job. In most cases they will be under stress, not only because they need a job doing, but also because they are about to invite a stranger into their home.

Here is a scenario. A woman wakes up and goes into the shower, only to discover she has no hot water. What does she do?

There are several options;

1. *She calls a friend and asks if they either, know what to do, or can recommend someone who knows what to do*
2. *Calls on a plumber she always uses and trusts*
3. *Looks in the local directory for a plumber*
4. *She boils a kettle and gets on with it, thinking she'll deal with it later*

Let's say she draws a blank on 1 and 2 and doesn't fancy leaving it till later. The only option is for her is to find someone.

Now she goes through the directory and finds a sea of plumbers all looking the same, until she comes across a female plumber. Not only that, but emblazoned across the ad are words like 'respect, peace of mind, aftercare, quality of work'

What would you do?

Doesn't that client deserve for the plumber to turn up? She has put her faith into calling on a stranger and inviting them into her home. The least she can expect is that the plumber turns up!

Will that be all, madam?

In my early days as a plumber I made it my mission to leave the place looking better for having me there and not worse. A particular case in point was a woman whose husband had just left her with two young children. I was working in her house for three days. It was obvious to me that she was finding it difficult to cope with her new situation. The tidy and well kept house I had visited a few weeks earlier to price up the job had turned into an upside down mess. To complete the job I had to go into each room of the house and it was clear she was not keeping on top of the household chores, especially in her children's rooms.

On the last day she and her children were out. I finished the job, put all the carpets back down and took a last look around. Then I decided to go just that little bit further. I went through the entire house and cleared up. I collected the rubbish into bin liners, I put the books onto shelves and made the beds. I loaded the rubbish into my van and vacuumed the carpet. It took me an extra hour to do the tidy up.

Later that evening I answered the phone to the customer who was so grateful she was crying on the end of the phone. I didn't charge for the extra but she paid me more anyway. To her, the simple fact that I had done more than she had asked was worth a great deal to her. It also became quite lucrative for me because she spread the word to her friends and they all used me.

It is very short-sighted of any business not to consider good customer service. It doesn't have to cost money or take much extra time. The rewards for giving just that little bit more are far reaching. The plumbers interviewed for this book, seemed to go that extra mile as standard. For them, as for me, those touches not only made a big difference to the customer, but also to us, the plumbers. Everybody likes to feel good about themselves, and every plumber I interviewed said, they wanted to see the customer happy, because it meant to them that they had done a good job. This is one of the reasons why customers love female plumbers.

Of course, this is not the exclusive domain of the female plumber and in the interest of fairness here is an example of a male trader who went the extra mile for me.

I had cause to call a plasterer to my own home some years ago. The job he did was pretty good, and I was happy. I would've recommended him just on that but what made me be sure to recommend him was that he spotted a patch of wall, that was not part of the job, but needed to have a finish of plaster on, and he did it. He didn't charge me for it, he didn't make a fuss about it, he simply got on with what needed doing.

Give a monkey's

I was once called to take a look at a decrepit bathroom that needed a total rehash. A beautiful Victorian bathroom had been replaced with a 70s 'champagne' suite, yellow plastic shower screen and brown tiles!

The clients couldn't wait to have it all gone. I was surprised when the first thing they said was 'I suppose the layout has to be the same' and 'whatever is easiest for you' I had to draw the line and say 'actually, it's whatever *you* want.'

When was it decided that the job had to suit the plumber? In my book, it was always down to the customer to decide what they wanted and me to deliver it. Good customer relations don't depend on the customer bending over backwards to please the plumber, rather the plumber should bend over backwards to please the customer, without compromising themselves or the customer.

My research showed that price, whilst important, was not the number one concern of customers. What they wanted most was to be able to trust the person coming into their home. If they could do that, the cost was far less relevant.

A common thread amongst the women I interviewed was that they wanted to please the customer. This means, paying close attention to what is required and delivering it efficiently and cost effectively.

There's more to it than that

For years the nature/nurture debate has raged on. Are we the product of our birth or of our upbringing?

My next-door neighbours have two very young children. They are a modern, liberal couple with a very rigorous ethical code. One sunny morning the oldest, a boy aged four at the time, quite matter-of-factly stated, 'you're not pretty' His parents were mortified and immediately insisted he apologise for his remarks. I assured them I was not offended (I'm actually quite pretty, as it happens) but wanted to know why he had said that. So I asked the parents to indulge me. When asked to explain he said I wasn't pretty because I wasn't wearing a dress, and 'pretty ladies wear dresses'. I explained that it would be difficult for me to 'plumb' in a dress, but showed him a picture of me in a dress to see if his opinion was different. It was, much to the relief of his parents, who pride themselves in the diversity they have shown their children.

Even though these children were brought up by broad-minded parents they still had conventional ideas that women should have long hair and wear dresses.

Plumbing may be seen as a male-dominated area, but as I have said before, it's not really about the plumbing.

It's the story behind the story. Customers love female plumbers for the same reasons they love female electricians, carpenters and insurance brokers.

They love us because we care

Consider this. You are called to a customer distressed because water is pouring through a ceiling. When you get there, the first thing you do is calm them down. Your calm ways and soothing manner makes them feel cared for, much like a mother would soothe a crying child. They feel rescued. Then you proceed to fix the problem.

Now let's unpack that a little further.

You've heard of Sigmund Freud, the 'father of psychoanalysis' and after who the well known phrase 'Freudian slip' is named. He's the one who said that all our problems stem from our repressed sexual desires.

Don't worry, I'm not going to make a sexual association with women and pipes. Although once I was introduced to an audience thus; 'This is Hattie and she is into women's plumbing'. It got a laugh, and in the break someone came tripping over to me thinking I was a gyneacologist!

Freud's young protégé, Carl Jung looked at ancient myths and archetypes. He described the human psyche as being drawn to certain archetypes throughout life. An archetype is the embodiment of a trait or behaviour. So, Superman is the archetypal hero, the Pope, (for some) is the archetypal father. Fairy stories are full of them and that is why they appeal to us as children and adults. Jung developed a whole new understanding of human behaviour on which all personality tests are based.

I'm not going to go into the theories of either of these men, there is plenty of material already in existence. Suffice to say Jung's archetype theory is very relevant in our exploration of why customers love female plumbers.

In 2009 I was extremely fortunate to spend a day with branding and archetype expert Helen Urwin. She was instrumental in developing such iconic brands as Mars, Sheba cat food, and Dolmio pasta sauces.

We were able to unpick the reasons why customers love female plumbers and add depth to Stopcocks as an iconic brand. In essence; when we turn up we are heroes, then our nurturing ways make us mother, then our ability to fix the problem makes us father. So we embody three crucial archetypes, which the client is unconsciously attracted to. We'll look at archetypes in more detail later.

It's probably true to say that when it comes to buying new things for the home, it's the woman of the house who takes charge. She decides what's needed and then she goes out and chooses it. (She may get someone else to pay for it).

When the house 'breaks' it is usually the woman of the house who notices and wants something done about it. Now I know it's not ALWAYS the case ......

but it USUALLY is the woman of the house that gets things done.

Some time ago, while I was working, the woman of the house took a shower. This is not the first time that has happened. I later asked her if she would have taken a shower with a male plumber in the house. She was mortified at the thought. She explained that she'd thought nothing of taking a shower while I was there. On another occasion I turned up for an appointment to find the customer out. When I rang her she told me that she had been delayed but that the front door was open and I could go straight in and start the work.The couple arrived some half an hour later. I could have run away with the entire house contents in that time, but here, again, a complete stranger was trusting me to be honest and honourable. On a different occasion a young mother breast fed her baby while I serviced the gas fire!

It is an interesting thing, and many times I have tried to explain what it is that attracts people to women in the trades. I personally always felt that I needed to be safe and secure in my home. If things are broken then I experience it as chaos.

I see my home as my sanctuary and haven from the world. I like it warm and clean and working to its best. When it isn't, then you can bet it won't be long till I'm not either.

Most often when I am called to a home, it's because something isn't working. The householder is usually in a distressed state.

I am constantly walking into other people's stressful situations.

My job is to calm them down, then repair the 'broken' house.

I see the link between the home and the psyche.

Women are traditionally seen as the home-makers. We may be plumbers, but that isn't to say that we are not home makers. In fact I built most of the home I live in.

Being aware of this may help you when you embark on your new life.

Customers are predisposed to love you

If all that evidence wasn't enough to convince you that customers will love you, let's just look at something else.

Why was it that during one of the worse global recessions in 2009/2010 Stopcocks and indeed the businesses of all the women I interviewed grew? When traders were slashing their prices to cut throat levels, women plumbers were thriving!

The recession didn't stop people spending altogether but it did make consumers more careful about what they were spending on. None of the plumbers I met in my research for this book felt the recession. They remained just as busy, and, in fact many got busier than ever.

A customer survey (my own empirical research) revealed that customers expect a better service from women. The expect that a woman will turn up on time, do a tidy job and clean up after herself. They expect a woman to care about their property as if it's their own.

Perhaps it's these expectations that make us work in the way we do. Perhaps it's the way that girls are brought up to be mothers and carers that makes us work this way. Perhaps we are wired to be caring and nurturing and it doesn't matter how we reinvent ourselves, our approach will always be the same.

For me it's true that I want people to like me, I want them to be happy with the job I've done, and I want them to talk about me to their friends. (In a good way, of course). These have been the drivers that have made Stopcocks a success.

I didn't know I was being archetypal in my behaviour and neither, I expect, do you, until now; but customers want someone they can trust, and someone who understands what's going on for them when their house is 'broken'. They don't want a clod-hopper who traipses all over their house and their feelings, with no care.

The secret is; customers don't know this until they are presented with the possibility of a female plumber. They are 'trained' to believe that a tradesperson will rip them off and that they have to be ultra guarded. They put on their psychic armour when the doorbell rings. When a woman answers the call for help, they can take off that armour.

Two myths to dispel

Myth I

Female plumbers only have female customers

This is the greatest myth of all and is simply not true. Every plumber I interviewed stated that their customer base comprised men, women, single or married, young and old, families and lone dwellers. As I pointed out above, it is usually the woman of the house who calls the plumber and deals with the aftermath but she is most often part of a couple or a family. In fact, research shows that 80% of all buying decisions are made by women. With 31million women in the UK and 25% of those supporting women run businesses, we have a starting market place in excess of 7 and a half MILLION customers!

Even if that myth was true, we'd have enough work to keep us going for many many years.

Then there are the male customers. I would categorize them in this way. All men fall into one of four groups.

The sexist bigot

No more needs to be said about him. He doesn't believe women could or should be plumbers and would never call one. So he is off our radar and good riddance!

The man who doesn't want another man in his house

There could be several reasons why this could be so. Perhaps he doesn't want a man to be in his house alone with his wife or girlfriend. Perhaps he doesn't want another man to see his place. Perhaps the only testosterone he wants in his house is his own (thank you very much). Perhaps he feels he should be able to do this kind of stuff and feels he would be ridiculed by a male plumber. Whatever his reasons, this is all good news for you, because he will be calling your number when he needs a plumber.

The man who wants a clean and tidy job done and his house treated with care

This man wants someone who will treat his house as if it were their own. He wants someone to listen to his requirements and carry them out with the utmost consideration to his property. He will call a woman plumber because he doesn't want to take chances.

The right-on man

He will call you because it is 'the right thing' to do, and 'well done for going against the grain' and 'good luck to you'.

Myth 2

Your customers will mainly be O.A.Ps

It is true that the elderly are very vulnerable and would welcome you. I would advise that you form alliances with the local Help the Aged in your area when you start your business. They are one sector of society that will keep you in work and help you build your confidence, and that is not to be sniffed at. They will not, however, supply the majority of your work. Don't worry, the rest of the world needs you too.

Where can I get the report from?

Extracts of the report can be found at the end of this book. I talked to plumbers both male and female UK-wide to implement some of the changes outlined in the report. On one occasion it was suggested that the plumber call the customer ahead of time to confirm an appointment before turning up (on time of course) Just that one simple act gained that plumber the 'Superman' factor, which had a great effect on his business.

On the next leg of our journey we will be looking at our bodies and asking; What makes our bodies so perfect for plumbing even though we don't have the physical size or strength of men?

Web links

The full report and information about how to book time with me to go through the report can be found on the website as well as stories about customers loving female plumbers on a downloadable audio that you can get from **www.joyofplumbing.co.uk/downloads**

CHAPTER SIX

You have the perfect body

Two baby elephants

It's not often we are told as women that we have the perfect body but that is where our journey takes us next.

For us, it really doesn't matter what we do. In one way or another it seems we never have the perfect body according to society.

You can probably guess that I categorically disagree with that point of view and take the absolute opposite view, that you do have the perfect body. So that's that and I can end the chapter here.... but I won't.

As women we are constantly being made aware and made self-conscious of the way we are. We are either too fat, too thin, too short or too tall. We are too old, too young, too plain, too tarty, too quiet, too loud, too intelligent (too clever for our own good), not intelligent enough.... You get the picture.

So in our lives we are plenty used to being criticized and scrutinised specially our bodies.

This is where it ends.

Consider this; Every square centimetre of our body has air pressure pushing down on it. This is atmospheric pressure. If we measured the amount of weight we are carrying just by walking around it would be similar to a couple of baby elephants. The reason why we can't pick up a couple of baby elephants is because we are already bearing that load, regardless of our physical size. We are used to this and we also have air at the same pressure inside our bodies, pushing out that prevent us from being crushed by the weight of the air exerting its pressure on us from outside. We feel lighter in a swimming pool because we are no longer bearing atmospheric pressure but the moment we get into an ultra-fast roller coaster at high speeds we feel heavier and heavier as G-Force or force more than gravity exerts its pressure onto us.

The point of all this is that if we actually realised how much we were doing as human beings we would recognise how incredibly strong we are. Now apply this same logic to the amount of 'stick' we get for pretty much everything in one way or another and realise how incredibly resilient you are.

Simply being used to something does not negate the work you've had to do to become used to it.

I was born the second child of five and the oldest girl. As I grew up I watched as my parents' expectations of my older brother were piled on. They clearly wanted what they thought was best and fitting for a boy to do. As my brother reached his teens it became clear that he had his own ideas about what he wanted and they somewhat clashed with what our parents had planned. They wanted him to go to university, he wanted to leave school early. This eventually led to huge disruptions in our family and he left the family home in a blaze of chaos.

Being two years younger than my brother I remember watching these events unfold and deciding that I would please my parents and go to university. How wrong could I be.

Not for girls

My parents' plan for me, was quite different. For me the plan was, leave school at 16, and enter an arranged marriage, have children and be forgotten to my own history and propagate the history of another bloodline. I had to get used to the fact that my life would be very different to my brothers' lives. My 'role' was to marry and bear children just like all the rest of the women in my family.

This was a terrible rude awakening for me. I began to see and feel the inequalities of our existences. I got used to it, but that is not to say it wasn't a very hard road to travel. Once I was used to it though I forgot the harshness and the hurt it caused to my young psyche.

I was extremely lucky. I came from survival stock. My own mother had endured plenty of hardship and heartbreak (as did all five of we children) while we were growing up, so I quickly realised that I could escape my 'planned' future and create one of my own by becoming educated.

I was able to use the strength I had built to get used to my situation, into something that would propel me to where I wanted to go. I was the first of my family to go to university, although it was a proud achievement for me, it took years before my parents would accept that my life was my own.

Perfect body

Realising you have the perfect body doesn't start with getting onto the scales. Like everything, it begins with seeing that your whole life is perfectly described and portrayed right there for everyone to see in your body. Accepting that is the first step.

You'll see on this part of the journey that you have the perfect body for being a plumber and I will show you how and why that is true. To recap; there are many things in our lives that we get used to and having got used to them we forget the learnings and hardships we have overcome on the way. We simply live our new lives adapted by what we have learned. If we could only realise that we worked damn hard to get used to those old situations, and build on the strengths we have accumulated, we can move our lives into any direction we desire. If you realise that you have worked damn hard to get used to your situation you will realise that you have amassed an entire toolkit that is yours and ready for the using.

A very good friend, pole dance tutor and pilates instructor Goddess Star Monroe writes, 'To be strong is to be feminine' You can read this and a collection of writings from other experts at the end of the book.

Strength is more than muscle

A woman heard a scream and saw a child had been run over by the ice cream van in the local neigbourhood. Without a second thought (probably, without a first thought) she grabbed hold of the van and hauled with all her might until the child was able to scramble out from beneath the wheel. Then she picked him up and carried him home to his own mother, who was able to take him to hospital to repair his mangled leg. He made a slow but full recovery.

This is a true story of heroism. The woman in question was my mother-in-law, who was 4ft 10 ½in tall and in her 60s when this happened. She had tapped into strength she couldn't have dreamed she had, and certainly couldn't access at will. She could only say that if she had stopped to think there would probably be no way she could have lifted the van. Where did this strength come from? How could an elderly woman harness this kind of strength? Obviously, there was something inside her she knew nothing about. Even into her 60s she was able to access it in an emergency, showing it was always there, lurking. This story tells me to never underestimate our own power.

We are so much stronger than we know and now that you realise you are bearing the weight of two baby elephants as well as all that hidden extra strength and that you have worked damn hard to be able to be used to your situation, I hope you feel stronger than you did 10 minutes ago. If you are still not convinced, try this simple exercise.

I'll begin the list of things that I have had to be strong about. These may or may not apply to you. I hope you will take a few minutes to add some things of your own. Think back to your schooldays. Think back to the things you just got used to, and then remember what it took for you to achieve that.

- Not speaking English as my first language
- Having no friends in nursery
- Being poor
- Not allowed to bring friends from school home
- Having to do all the household chores
- Violent father

When you've finished, take a look back and realise just how resilient you have learned to be.

All that AND two baby elephants takes a lot of strength.

So now you're feeling stronger than ever I hope. But how much physical strength do you really need to be a plumber?

One of the constant questions I'm being asked by non plumbers is, 'how do you cope with all the lifting, isn't it a really heavy job?' There are some heavy aspects to plumbing. For one, some of the tools are heavy and often you need to carry a heavy toolbox or ladders up and down stairs. Your toolbox will probably weigh the same as your toddler. This would never prevent you from picking up your child though would it? If you carried a power core drill around with you for three months without putting it down it would be equivalent to the last three months of your pregnancy. (Probably considerably easier on your bladder though).

Health and safety legislation in the UK guidelines state that weights of more than 35lbs (2½st) be lifted by two people. (the guides for men are 55lbs). That means that when my golden retriever came with me to work one day and was frightened of coming down the stairs, I should have got someone else to help me carry her because she weighed 45lbs. Next time your six-year-old falls over, call on a neighbour to help pick them up.

This is ridiculous but we do it.

Now it is true that sometimes we are called to lift heavier weights, some wall hung boilers, long radiators or cast iron baths weigh far in excess of the guidelines for both men and women but when you find yourself in a situation where you are on a job on your own with long awkward radiators to hang there is often very little option but to 'get on with it'.

In most cases, weight is not the issue. Our bodies are extremely adaptable and before long you will be carrying your toolbox with ease. If you have been plumbing for a while you will have noticed your body growing stronger. If you are about to start and are concerned about the heaviness of the job, then take it from me; you will quickly find your back, arms and shoulders building strength.

Being generally less muscular gives women an advantage

I was asked once about whether women have the strength to tighten things up enough. The truth is, over-tightening is more often an issue. Tightening a compression fitting too much can break the seal created by the olive, becoming too squashed. Rubber washers can tear and stretch if they are tightened against too much too. This is something that is taught at college, but soon forgotten once out in the field.

Over the years I have certainly found myself in situations where I have had to move heavy objects. Being only 5ft2in and slight never stopped me from hanging a boiler on a wall 8ft high or moving a cast iron radiator.

A customer was astounded when he returned home to find his new boiler hung so high. When he asked how I'd managed it I answered 'with my brain'.(The full story can be found later) Being of small stature and build I have always had to find different ways of working. Not having the bulk or strength of a man meant that I instinctively worked in ways that would preserve and protect me from injury.

Sometimes the best way to protect myself from injury was to call for help. I have on occasion knocked at the house next door to say 'I need a man'. This has always been a useful strategy, because men like to be called upon for their strength.

Plumbing is not so much heavy as awkward. A fibre glass bath may be light in itself but the weight is distributed in odd ways, making it awkward to move. Boilers are heavy on occasion, but not prohibitively so. Radiators have hand holds and don't need lifting too high to be hung. A 125mm core drill for cutting a perfect hole for a WC or flue can be heavy and hard but frequent rests or finding different ways to hold the weights gets the job done.

In a barn conversion in the middle of the Yorkshire moors I spent the entire day drilling 20 holes of various sizes through 24in walls for the five bathrooms, two shower rooms a utility room and the kitchen! Most of the time holding the drill above my head and standing on a platform or ladder. Yes, that was very hard. A long, long hot shower was needed when I got home on that day but the customer was happy with the job and my standards, my working methods did not concern him in the slightest (in fact the customer was a local electrician I had worked with, and his family; he already knew how inventive I am as we'd worked together on site).

The point is well made that even though the job is heavy at times it has never prevented me from getting it done. Similarly so for all the women I interviewed. Each of us have found a way of working that suits us. Which includes finding ways of managing the occasional heavy aspects of the job.

Much more common and where being smaller is really an advantage is the fiddly-ness of plumbing. There is far more need for delicate handling in plumbing than you can at first imagine. When you get to repairing gas fires and boilers, it's more like being a motor mechanic (without the oil). Here, nimble and small fingers are needed to get to those awkward microswitches and burner jets.

Bathrooms come in all configurations. Layouts are designed to look good not to be easy for the installer. Bath taps are particularly and notoriously difficult to get to. Often you will need to contort yourself into the most awkward positions to gain access. In the UK, bathrooms were brought inside in the 60s and 70s which meant they were often crammed into small spaces. We are ideally built to get into these places and I have been known to get completely underneath a bath to fix it to the wall in an inaccessible corner!

Leverage

Here is a story of perseverance, persistence and leverage. Many plumbers work alone, and that is the same for women plumbers too. I have found myself in situations when I had no choice but work out what to do and do it. A case in point was a small house I was working in where I was ripping out an old gas fire and cylinder, to be replaced by a new cylinder and central heating system. It was a total renovation project so the house was unoccupied which meant I could have free reign with the water. I had turned it off and drained down the cold water storage cistern ready to begin. When I removed the gas fire I was surprised to find a small iron back boiler with two lead pipes coming out of it. I assumed this had been long out of use and proceeded to cut through one of the lead pipes. As soon as I made it through the thick wall of the pipe it began to seep water. This didn't worry me and I went to fetch a bucket. I continued to cut through believing that the water would stop long before the bucket was full. (You know where I'm going with this don't you?).

As the water level in the bucket got past half way I began to suspect that it perhaps may not stop, so I went in search of another bucket, in vain. I was, by now getting concerned as the water level was rising and I was not able to find a way to contain it or stop it at this point. There was really no harm to be done. The house was completely gutted and the water would not have caused any damage, but I was confused as to where it was coming from. Perhaps you have already guessed, this was a primatic cylinder. This is a type of hot water cylinder, where the heating circuit and the domestic hot water are separated by an air bubble within the cylinder itself. When I cut the pipe and started draining it down, this air gap was broken and the entire cylinder was emptying into my one bucket!

Of course, I didn't have time to stand around and theorise about this, I was too busy trying to figure out what I as going to do about the rising water level.

I needed to think fast, so I scouted around the site. I wanted to be able to control the flow in some way. I found a small plastic lid amongst the rubble and wondered how I could use it.

Lead pipes are very heavy and I wanted to use this to my advantage. I lifted up the lead pipe and bent it towards the wall. Placing the plastic lid over the end and using the weight of the pipe to hold itself jammed against the wall to stem the flow. That was one end of the cut pipe sorted, but being a circuit meant that water was coming out of both cut ends, so my task was only half done. Luckily, the water had no great pressure as it was only coming from the room directly above so I was able to knock a piece of wood into the end to slow the flow enough for me to empty the bucket. With that done I had more time to fashion a method of controlling the flow. At this point I still didn't know when or if it would ever stop. (At least if I could control it, I could figure out what was going on). While the bucket was filling at a slower pace and for the second time I quickly put together a washing machine tap and hose onto a short piece of 15mm copper. I was able to push this into the malleable open end of the lead pipe and turn the flow on and off at will.

Now I had time to ponder on where it was coming from and whether it would ever stop. With the flow under control, it was easy to limit the damage. The cylinder was soon empty, and I was able to get on.

I learned a valuable lesson that day. I learned how resilient I was and about the standards of work I hold. The water from the cylinder would not have been damaging and I could have let it fall onto the stone floor. It would have dissipated with no ill effect to the building, but that is not how I worked.

For me, and for the women I interviewed for this book, it is simply not an option to allow water to fall out of a pipe uncontrolled.

Here, some knowledge about the material (the heaviness and malleability of the lead), persistence in finding a solution, and perseverance meant I was able to keep to my standards. It felt pretty good too, especially when I went upstairs to find an empty cylinder that was easily carried downstairs.

On another occasion, the boiler was to be hung inside a cupboard at great height. The boiler itself was not too heavy for my, by now, developed muscles. I used a series of levers and platforms to get the boiler onto a chair, then onto the washing machine. Then I stood on the chair to lift the boiler into the cupboard. (It was an alcove cupboard made of solid timber and could take the weight). Then I stood on the washing machine to lift the boiler onto its brackets.

Here, I used the fact that I could lift the boiler to waist height to my advantage. I didn't consider it an impossible task, just because I couldn't lift the boiler up above my head.

A third example of leverage was used by myself and my brother-in-law to carry a cast iron bath downstairs in one piece. We set up a series of pulleys and ropes and instead of breaking our backs (and possibly the house too) we simply needed to control the positioning of the weight as the bath was gently lowered down the stairs and out of the house.

Perhaps one of the most useful 'tools' in every female plumber's toolkit is her brain and her ability to use her body in inventive ways to get the job done. I don't doubt you have come across many situations where you have had to think on your feet, and in your plumbing career, that will happen on a regular basis. This is character-building stuff. As your body grows stronger, so your confidence will grow too.

You will be asking a lot of your body, especially your back, arms and knees. My advice is this; take care of your body and it will take care of you. Build up your strength, protect yourself from doing damage to your back and knees. Invest in proper knee-pads. Stand firm and keep your centre of gravity low. Always bend your knees when lifting, and when necessary ask for help. There is no shame in asking for help. There is no gain in trying to do everything by yourself and risking injury.

Big strong woman

I hope by now you are convinced you don't have to be a muscle-woman to be a plumber. I hope you can see that you have the perfect body for the job. I hope you have realised that you don't need muscle alone to be strong.

Women are so resilient. You are so resilient. It is in our make up (excuse the pun) to be resilient. All About Brands director of talent Bridget Biggar has written a piece about the resilience of women that can be found at the end of this book.

There is another aspect to consider.

You are not alone. You don't have to be a big strong woman doing everything by yourself. You may be the only female student in your class, or in your area but you no longer have to be isolated. Since the beginning of 2010 Stopcocks has been building an international community of female plumbers. This community includes existing successful plumbers, student plumbers and those who are finishing their qualifications and coming out into the world.

From now on you can join, search and find other women plumbers at the click of a mouse.

Isolation is a big issue for female traders. The online community was set up for you and it has everything you may need from finding the right insurances for your business, to getting your admin done, to buying materials at the best prices.

You are not alone, there is a whole community of big strong women out there waiting to hear from you.

See you there.

If you are ready to move on, then, on the next part of this journey we will be exploring the genders. The only part of the journey that directly compares men to women and far from what you would expect, this is a celebration of gender differences. There is no doubt that women need men and men need women. There is a good reason why we are different and it has nothing to do with what jobs we decide to do.

Perfect Body

Web links

You can download my guide to lifting by going to **www.thejoyofplumbing.co.uk/downloads**.

You can join the Stopcocks online community now by going to **www.stopcocks.co.uk**.

Download the interviews from **www.joyofplumbing.co.uk**

CHAPTER SEVEN

Let's talk about sex, baby

Nature or nurture?

In an experiment in the 60s to see whether boys and girls were treated differently three women sat on three benches in three busy shopping malls, each with a baby in a pram. The difference was, one baby was dressed in blue, one in pink and the third in yellow. All romper suits exactly the same except for the colour. The women were instructed not to reveal the gender of the baby unless specifically asked, and not to correct the passers-by at all. The experiment was to gauge the reaction to the babies by the passing shoppers. The results were both revealing and shocking.

Very few of the passers-by asked the gender of the babies, their reactions indicated that the genders were assumed.

> The babies in blue and pink evoked reactions that alluded to their gender.
>
> The blue babies got 'aren't you a handsome boy'.
>
> The pink babies got 'ooh what a pretty girl'.

The yellow babies however did not evoke such reactions. In fact, the yellow babies were ignored significantly by the shoppers.

It appears that the reaction to the babies was influenced by something that led the shoppers to believe something about the babies to be true. In this case, the colour of the romper suits seemed to indicate a specific gender, and therefore created the reactions by the passers-by.

(Don't worry; I do believe the yellow babies were not harmed by this experience).

I find it incredible though, that so many reactions are just automatic. The blue and pink phenomenon is especially interesting to me, particularly since I, like you, am a woman doing a job that is not traditionally meant for me to do.

My upbringing certainly differed from both my brothers' even though one is younger. However the blue and pink phenomenon was never present in our home.

Being elected as a captain for a school group during a visit to the Science Museum I elected to call our group 'Orange Group', much to the dismay of all the boys who were all rooting for 'Blue Group'. I didn't see what all the fuss was about; We only had to wear an orange sticker.

My story of discovery

Although my father was an electrician I was 11 years old before we got electricity put into our bedroom. Before that, my sisters and I shared a room that had a lightbulb on an extension cord from my parents room. When it was lights out my father would simply turn off the light switch and we would be plunged into darkness. The good thing about this was that we made the best out of the time we had before the lights went out. I was always a prolific reader which (unfortunately) was not encouraged for girls in my (Turkish Cypriot) culture but it did mean that I read everything going, not only to myself but to my young sisters too.

When I got the chance I would be taking things apart and putting them back together again, to see how they worked, (again, not really what I should have been doing).

I had been given a pocket radio that ran on batteries but I didn't want to spend my pocket money on buying the batteries that used to run out so quickly. I knew that batteries were a source of electricity so I theorised that the radio could be powered by our extension cord. I took the light bulb and fitting off, (after turning off the light switch in my parents room first, of course) then I connected the two wires to the two terminals of the battery connector. I did this using scissors and Sellotape. Can you guess what happened when I turned the light switch back on?

Yes! There was an almighty bang. The radio shot up into the air, hit the ceiling and landed on my bed in a smoky, smelly mess and the house was plunged into darkness and silence as the entire fuse box had switched itself off. I realized the difference between AC and DC (kind of) that day but it didn't stop my curiosity. When electricity was brought to our bedroom some years later I was the one that went under the floorboards to pass the cables. Perhaps my interest in engineering came about because of my father's occupation but it certainly became very useful, when my father was absent and anything needed doing in the house.

This was all very well, until I began to sprout tiny bosoms. The fun and games ended there, and the serious business of finding me a husband began.

The case was very different for my brothers. They were encouraged to go out, and see the world. They were taught to be dominant and domineering. This was how 'men' are in my culture, and it certainly was how my brothers were.

So I began to plot my future from the age of 11. It was clear that my life would be very different to most, if not all, of my peers.

What is a tomboy?

I was what you would traditionally call a 'tomboy'. That is, I liked doing things that boys did. I liked to take things apart, ride bicycles, play football. My younger brother and I had a favourite role-playing game where we were both spies (I was a male spy too). Everything around me seemed to look better for boys. They had the best toys, the best clothes, the best games, the best life. So, when I compared my life to that of my brothers, it seemed to me that being a boy was the best and being a girl was naff.

I wanted to understand why we were different from a very early age. I devoured books and articles about the gender wars. There was one crucial element that I was missing as my young adolescent brain was developing. Being a boy or a girl has nothing to do with the things we like to do. All of that stuff is taught us from a very early age.

I associated a good, fun life with the life my brothers had and I didn't. So I wanted to do those things and have a fun life. I even thought I didn't like being a girl, because of those reasons, and wanted to be a boy. My understanding grew when I studied psychology at university. I began to see that the difference between male and female was not the same as the difference between masculine and feminine. This seems to me to be a common misunderstanding, that leads society to say that women can't (or shouldn't) be plumbers and men can't (or shouldn't) be nursery nurses.

I began to see myself as a woman who liked to do things for myself. I rode a motorcycle and became one of the very few female motorcycle despatch riders in London. I repaired my own bike and eventually when I got my first car, I was able to do minor repairs on that too. When I look back on those times with hindsight, I realise that it didn't matter what I did, my approach to it was always feminine. I always thought like a woman and felt like a woman. Even when I delivered a package, it was in a womanly way. The fact that I did motorcycle repairs had nothing to do with my gender. This was very liberating to me and made me see that I could do and be whatever I wanted.

What makes us so different to men?

So OK, we've looked at the physical differences. We know that women are built to put on fat, so they can incubate and feed offspring and men are built to grow muscles so that they can do the other stuff, that isn't having babies.

There are plenty of studies around showing that women and men hold conversations differently. Women, tend to look at each other more in conversation and tend to want to know more details, where men don't look at each other as much and are more focused on concluding a conversation with a result and moving on.

Two women talking will go down tangent after tangent filling in more and more detail, where as two men will get straight to the point, go through the resolution and move on to the next point.

This is usually the case, although I am reminded of my first week at plumbing college. I joined the guys for lunch in the cafeteria. I was new and it was obvious they were all familiar with each other. Imagine my shock when I overheard the conversation that was going on.

Two of the guys were talking about eating a healthy diet and one was relating how he had recently started having muesli for breakfast rather than the usual fry-up to try get his cholesterol down. Another was talking about picking his young child up from school and wondering what to give him for his tea.

The third conversation really made me feel like I was in a parallel universe. I overheard one of the men saying he had to buy some new clothes for going to his wife's parents' house for Sunday lunch. He was asking his friend whether he thought he should go for casual or smart, and what colour he should wear!

This really opened my eyes. Not one football conversation or obscene joke to be heard. Had I got it all wrong about men? I realized I hadn't got it totally wrong when soon after when I sat amongst another group, this time of younger men. The football, obscene joking and bravado was out in force and I wondered if these younger guys felt they perhaps had something to prove. Maybe with family and responsibility comes compassion and caring and the confidence to be who they are rather than what is expected of them. Perhaps there is room for both compassion and 'blokey-ness'

We can't live without them and they can't live without us

During the 80s I had the mixed blessing of living in a separatist household. What on earth is that? I hear you ask. The 80s was a very politicised decade in the UK. Women had a loud voice in many of the issues of the time. There were strong movements against sex discrimination in the workplace, a huge anti-war campaign was staged by women and led to the famous Greenham Common Peace Camp where a number of women lived for months, some for several years.

Anti-fur and animal rights protesters were largely women and many women lived in houses that were strictly no-go areas for men or boys. I lived in one of those.

I say it was a mixed blessing, because it gave me insights into why it was important for women to have these places where they can just be amongst other women. It also gave me insights into why this could not be a way of life for me. It is why I can write with certainty; women need men and men need women.

Let's take a look back at prehistory. When early humans roamed the earth, before agriculture created settlements. Humans were nomadic, they lived a life dictated by what they could hunt, kill and eat, and what they could forage for. These hunter-gatherers lived quite successfully. The females were revered, respected and protected as they carried and bore children.

There is no evidence to show that there were any differences in the roles taken by the genders, but a woman with child was obviously unable to hunt and had to be taken care of and protected during her most vulnerable times. It would probably not be unusual for several women to be pregnant at the same time, and need protecting, not only while she gave birth but also while the children were young and certainly, until they could walk and run for themselves.

While these tasks were performed by the women, the men would be out hunting and gathering to feed the tribe.

Could the present day gender roles be a throw back to our very earliest pre historic times?

Are we still prehistoric?

When I first started as a plumber I noticed that the plumbers' merchants and builders' merchants always had men just 'hanging about' I would rush in, buy what I needed and rush out but there would be guys standing around as if they had nothing to do.

I wondered if they would just go there to socialise, or they had friends there they liked to hang out with. None of this was true. In my travels around the world (Asia, The Far East, Africa, India as well as the Middle East and USA) I was astounded to find that men all over the world would be 'hanging out', while the women carried the water, built the roads, did the shopping, cared for the children.

I started to wonder about why this is. Why are women always moving and doing, and men sitting around? I tried it once. I tried to sit around and do nothing. I went to my local riverside and went fishing. I actually enjoyed the peace and quiet but I used the time to plan my next day and I could only stay until that process was over, then I had to leave. I managed three hours but only because I forced myself to. A whole hour of that I spent telling myself that I didn't have to leave. After an hour of fighting with myself, I gave in and 'got on with something.'

So why do they do this? Maybe we are misunderstanding them. Could it be that in our society now, where there is no need to protect against wild animals, or neighbouring tribes the male role as a protector of the homestead has been usurped? Do they not know what to do with their time, are they just lazy?

There is no way I can know the answer to this question really, but it did help me to adapt my expectations.

What about equality?

I have listened to many discussions, read many books and seen many people talk about equality. It has even been made into a law. But what does equal opportunities really mean? In most of my experience it has meant that we are all treated the same. The same as what, though?

Any man in a predominantly woman's world will tell you that they are still treated like men. What about women in a man's world? We are treated like men too. I recall the tutors at college, making it quite plain that they would not discriminate against any female and they would treat us all the same and continue with their swearing and belittling as if gender was no issue to them.

I would like to take issue with this though. Think about this for a moment. Would it be a wild and crazy world if equal opportunities meant . . . wait for it and can I have a drum-roll please . . .

everybody was treated like women

Is that too bizarre a world to contemplate?

For years, to get ahead in business, a woman had to emulate a man. To a large extent she still has to keep her emotions in check, wear masculine clothes and grow some (albeit proverbial) balls. The effects this has on women is well documented and you can read about it at the end of the book from an authority on the Essence of Womanhood Susie Heath.

Away from the workplace is another story. It's all too easy to talk about equality meaning that men should wash up and women should mend the car. Those things could certainly be a by-product of equality, but they are not what makes up equality.

Society is learning all the time from its mistakes. From positive discrimination to the whole PC (politically correct) phase of the 90s (which has become a joke), no actual difference has been made to society in the slightest. Why?

Because you cannot legislate equality.

Equality is a state of mind and a state of being. How 'equal' do you feel in your life, your job, your relationships? I don't mind how many times I wash up, or mend the plumbing so long as I feel equal. It has nothing to do with what job I do, or what tasks I perform but has everything to do with the way I feel about my situation.

How do you feel about your situation? Do you feel strong at work, or do you feel down-beaten? Do you feel heard, or ignored in your family life? Are your relationships built on true acceptance of your partner who truly accepts you, or have you settled for always doing what he/she wants and never speaking what you want?

These are all indications of equality, not whether he washes up or not.

Where do we get this equality from?

In Chapter 3 we delved deeper into confidence and, yes, you've guessed it, our feelings of equality come from ourselves. We are so extremely lucky in the developed world to have the luxury to grow our confidence.

Before we go on I would like to acknowledge where we are at and state that this is a luxury and even if you are feeling pretty downtrodden in your life, it is very important and a crucial step to get your life into perspective. When we feel lucky, we feel grateful, and when we feel grateful, we create our own luck.

There are women and girls all over the developing world who will never get the opportunity to ponder these questions as we do, and as I am asking you to do now. I invite you to watch a most inspiring talk by going to **http://www.ted.com/talks/lang/eng/eve_ensler_embrace_your_inner_girl.html**

Here Eve Ensler (Playwright and creator of The Vagina Monologues), tells us about some of those women and the strengths they have found even though they have suffered in the most appalling ways.

I hope you took the time to watch the film and have come back with a new perspective ready to start creating a life for yourself where you realise that it is not men, or law, or society that perpetuates your feelings of inequality. It is within you to become equal. To change your state of mind and find the confidence within you, by being bold enough to state that you deserve it, and taking those first steps. And no I certainly don't mean become like a man. I mean; to echo Eve Ensler, free the girl within you and allow even your menfolk to access theirs. (More on that later).

You are not alone. There are many women doing this right now and I talk and meet them every day*.

So what about our differences?

All cultures have folk tales or fairy stories that have been passed down by word of mouth from generation to generation. Red Riding Hood, Cinderella and Sleeping Beauty to name but three.

Now let's break those down and take a closer look. They have a few things in common. They have a leading lady, be she a girl or a woman. They have a hero usually a male character, and they have a baddie in the shape of a beast, witch or step-sister.

Carl Jung spent much of his later life studying the characters of folklore and fairy tales and postulated that these characters and characteristics survive in different forms all over the world for so long because the behaviours are universal. He connected these 'archetypes' with his personality theories. It is no mistake that these characters are so extreme. They represent our deepest wants, needs and fears. We love it when there is a happy ending. We want good to defeat evil. We prefer love to hate.

Put simply, our instincts are working all the time, under the surface. We react to things in ways we cannot immediately understand. In our lives we are so busy, so distracted by our jobs, our children, the TV news, that we barely have time to understand ourselves at all.

Archetypes – What are they and why are they so important?

This is an extremely simplified version of over a century's work on personality structure and relationships. I have simply picked out the pieces that are relevant to our journey. If you wish to delve deeper into psychological theories you can find much more detailed information about the archetypes on the internet.

The archetypes are important to our journey here. They give us insight into the difference in our approach from that of our male counterparts.

As I've already mentioned, all cultures have their versions of folk tales, myths and legends. These characters are the embodiment of universal human traits.

The classic feminine archetypes are maiden, mother, wise old woman.

They embody the three stages of womanhood. Each have their own unique characteristics.

The maiden is innocent and trusting. The mother is nurturing and loving. The wise old woman is healing.

The main male archetypes are hero, father, sage. The hero swoops in and saves the day, the father is the male care giver. The sage is the wise man.

The female archetypes are receptive, and all-seeing, which means they have a holistic view. The male archetypes set boundaries.

In men and women these manifest themselves as the masculine solution-focused behaviour and the feminine people-focused behaviour. So the man will come in, see the problem and fix it. The woman will come in, see the person as well as the problem, and then fix it. Both are valid and valuable traits to get the job done.

We are constantly being told our emotions are our downfall. Men are being told they need to express themselves more. It seems neither sex can win in this. But these fundamental differences are what make our approach to a problem so varied. As women we feel an emotional attachment to our customer and their requirements. When we cross between the female and male archetypal behaviours we appear to be almost superhuman.

That is not to say that men don't do that too. Those men that leave 'macho' behind and cross between the gender archetypes look superhuman too.

Archetypes get away with anything

This has been confirmed to me on countless occasions.

I was called one afternoon by a damp-proofing company. They were finishing off a job but had disturbed a mains water pipe which was then gushing water into the cellar. I arrived at 5.30pm to find a street full of damp proofing vans, four damp-proofing men and three householders all staring at the new water feature.

The relief was palpable as I drove up. I assessed the situation and confirmed that there was no stoptap in the building at all. The only thing to do was turn the mains off at the end of the terrace which would cut the water off to 15 other properties. I instructed the damp-proofers to knock on the doors and inform the neighbours the water would be off and gave them all 10 minutes' notice before turning the tap. Now I was able to do a proper assessment. While the damp proofers bailed out the soggy cellar I went about creating the temporary repair which would mean the water could be turned back on and everybody could get on with their evenings. This involved replacing an old lead pipe across two terraces and fitting a new stoptap inside the newly damp proofed house. By this time the whole street was outside watching to see if I could do the repair before it got completely dark. I worked on steadily and received a rousing round of applause when the water was turned back on.

I returned the next day to fit the rest of the pipe work.

Behaving as the archetypal hero, mother and father, I was able to deprive the whole street of water and disrupt three properties for most of the evening and still be thanked for it (and paid too).

As women plumbers we have somehow crossed over into the realms of both male and female archetypal roles. We enter a situation as a hero and care giver (mother and father) seeing it in three dimensions. Namely; the problem, the solution and the effect it's having on the householder. This gives us a unique perspective.

Our first priority is to calm the (possibly) distressed person. That done we neatly slide across into a male archetype and repair the damage.

Without being consciously aware of it we have interacted with the customer on a much deeper level.

Talking with customers (and female plumbers) over the past 20 years has given me a wealth of stories of how I (and you) provide such a different experience for the customer than male plumbers.

Studies show that men are more solution-focused. They enter a situation looking at the problem and finding a solution. They are just as competent (the good ones, that is) at repairing the leak as we are, and on many occasions no more is needed. Those male plumbers who show their caring side are the ones that are beloved by their customers.

So you can see we are different, in our physicality and our approach, and that it really has nothing to do with what jobs we do. I firmly believe, as people we can achieve anything we set our minds to.

We're about ready to move to the next part of this journey. Knowing how resilient you are, and what standards you set for your life are very important factors. Those things inform how you conduct your life and business. Times have changed. Business is no longer the domain of stuffy men in suits smoking cigars in the drawing room. The next chapter looks at the modern day business. Why today's way of running a business suits women particularly well.

Close your toolbox and come along with me as we look at you, your business and your lifestyle.

*I invite you to contact me **www.joyofplumbing.co.uk**.

CHAPTER EIGHT

Flexibility and Good Business

How it used to be

One of my favourite TV programmes when I was young(er) was The Fall and Rise of Reginald Perrin. It was the story of a man driven increasingly mad by the pointlessness of his job as a middle-class middle manager at a dessert factory. Told in a very comedic and humourous way and starring one of the UK's funniest actors Leonard Rossiter (star of Rising Damp) As the story progresses Reggie gets more and more eccentric until he is driven to faking his own suicide by stripping naked and walking into the sea.

The setting was very typical of how business was run in the 70s. The owner was a rich upper-class man usually only known by his initials (C.J), the managers were middle class, and of course the workers were working class. It was very clear where you were placed and very difficult to get out of your stereotypical place. You have to bear in mind also, that the status of women was below the status of men. So the lowest of the low, were working-class women. These women did the drudgery. They worked in the service industries, in factories and as cleaners and cooks. These were the lowest paid jobs but most essential to keep the country going.

The Women's Movement of the 60s and 70s had begun to have an impact but business was strictly the domain of stuffy men in suits. As women moved into the workplace it became clear that they would not be regarded as equal to men, even when they were doing jobs of equal status.

In my research for this book I looked for evidence of women in business in the 70s and came up short. Even in the US it seems that the position of women was still very much the homemaker and stay-at-home mother. This was well reflected in the TV shows of the time. The US Brady Bunch and the aforementioned Reggie Perrin. Even though there were more women in the workplace than ever before, it was just not done for them to demand anything but 'pin money', just enough for them to buy their own personal effects. The job of providing enough income to pay the household expenses was still down to the man.

A slice of my own personal philosophy here is; wasn't it incredibly short-sighted to stop women earning money? Wouldn't it have been better for the household if there was more money coming in? Surely it would lead to a more luxurious lifestyle if income was pooled.

There were great battles to be fought and won before women's pay was to come close to men's pay. The Equal Pay Act was passed in 1970 in the UK, but despite that the difference between pay for men and women is still at 16.4%.

What does flexibility mean to you?

The dictionary definitions are:

> **able to be easily modified to respond to altered circumstances or conditions and (of a person) ready and able to change so as to adapt to different circumstances**

Do either or both of those definitions fit you? Are you easily able to modify yourself and your behaviour to respond to altered circumstances? Are you ready and able to change to adapt?

My guess is that you are. I imagine that as a mistress of reinvention you have been doing just that for most of your life. You are probably so used to it, you don't even recognize it as a skill or strength. If this has not been added to your 'used to it' list, then do go ahead and put it on there now.

It's no accident that the female of the species is so flexible. Having to bear children is not without its difficulties and if there was no flexibility our species would not have become the dominant one on this planet. The human race as a whole is extremely adaptable and can live in most places on the planet but as we discovered in the previous chapter both genders have their roles to play and feminine flexibility is one of them.

When a woman gives birth the two halves of her pelvis split apart to enable the baby's head to pass through, both protecting the head from being crushed by the pelvic muscles and protecting the mother by holding her pelvic region together while she gives birth. Not only is the female body perfect for plumbing, it's also stunningly well adapted for having babies. The strength required to bear that amount of pain and do the extremely hard physical work of actual childbirth at the same time must never be underestimated. As I told a newspaper journalist,

> **'Anyone capable of giving birth to a child is stronger than they think.'**

So women are physically flexible in unique ways and it is this ability that also manifests itself in our everyday lives. In our reinventions we are constantly changing so as to adapt to different circumstances. Just like it says in the dictionary.

How many women do you know that 'put up with' things in their lives? How many of your female friends have settled for less than they deserve. They have adapted to live a life expected of them. Most women who end up in that kind of life have been so flexible they have forgotten who they are. We all deserve to have the life we want. There is no grand instruction book that says what kind of life you should live, (although if you have religious beliefs then you could consider your Holy Book to be the manual, and that is also your choice).

Does flexibility mean to be all things to all people and sacrifice your own needs/wants? Perhaps it could mean be all things to all people within the boundaries of being yourself.

One of the traps I fell into when I first started being my own boss was that I would try to answer every question and do everything the customer wanted. Even if that thing was prohibitively expensive, I would try to do it within the set budget. This was a big mistake. I lost a lot of money as I always tried to take up the slack, and that is not good business. As I gained experience I realised that most customers want your guidance. They have an idea in mind and a budget in mind. If those things are a million miles apart then they are depending on you to tell them that.

One such example is a cottage I was working on in the Pennine foothills. The customer had bought a derelict building and was restoring it. I had been called to fit the pipe work for the water supplies, bathroom and heating. The chaps had a pretty good idea what they wanted and what their budget was. They had a grand picture in their minds of the end product but no concept of what it would take to make their dream come true. When I arrived the place was an outer shell with no walls and only just a roof. This was an exciting project for me so I was keen to please the customers and give them whatever they wanted. My naivety led me to agree to coming in at their budget. The job took weeks longer than expected and I ended up seriously out of pocket. This is a lesson I learned the hard way. I adapted the way I ran my business from then on. I stopped giving quotes and only gave estimates, explaining that this would protect both the customer and myself from unforeseen circumstances. I built flexibility into my business practice.

As you gain experience you will learn that customers are looking to you for advice, and you will be able to give them that without compromising yourself too much.

One of the biggest calls on your flexibility will come when you have children to consider. This is when having your own business can really work in your favour. Jobs are notoriously un-flexible. It seems they still haven't evolved from when only men went out to work, and very few (even those that are predominantly taken by women) are flexible enough to allow women to work AND be a mother to their children. When you are own boss you really can decide that your day will end in good time to pick them up from school.

Many women (and you may be in this position) have been employed, had family issues, and been forced to choose between the two. I have spoken to a few plumbers where they have gone into employment as a plumber but had to leave because the job wasn't flexible enough to accommodate them. Unfortunately, these plumbers have been lost back to their original jobs. It's a common story and one that Stopcocks has been and still is determined to tackle head on.

Being your own boss

What does being your own boss mean to you?

For me it meant having the choice to do what I liked, when I liked. It represented control over my own life. Days off when I wanted, holidays when I wanted etc, etc.

The truth was that I did get to do what I wanted, when I wanted but not immediately. The reality was that I turned out to be the hardest boss of all. I worked myself to the bone, until I realised that I was being harsh on myself. I had lost sight of what being my own boss meant to me. When the work came in (and as I have said, there was no shortage of work) I took it. For the first six months I said yes to everything, and on many occasions I worked a seven-day week.

After six months I cut back my week to six days. After nine months I was doing five days. After the first year, I was so confident that I wouldn't be without work that I began to take days off. I would book myself a month in advance with days off whenever I wanted. I considered myself as a customer and booked myself into my diary. This gave me flexibility in a number of ways. If a job overran I had extra time but better still; if I finished a job early, I could have more time to myself and still know my diary was full. On occasion I would call a customer and bring a job forward, which was great for them and I could still have my time off.

I must also point out that there were occasions when work was scarce. In the beginning, I worried that work would dry up so I took everything and saved as much money as I could. If I had done a 10-day stint I could be relaxed about the phone not ringing for a couple of days. The phone always did ring. The work always did come in. This is also the story from all the women interviewed for this book.

It's as simple as you make it

In Chapter 3 I told you how I started my own business. A bunch of handwritten photocopied leaflets, some second hand tools, an answering machine and an A-Z. I also told you I knew nothing about business or marketing. I just knew I wanted to be my own boss. It really was that simple for me. Simple, but not easy. I knew nothing about tax and in the second year I had to pay back tax (I still have only a limited knowledge) I knew I had to keep records of my spending and, being me, I did put my receipts in labelled envelopes for each month.

I've spoken to many plumbing students and when I ask them about going to work for themselves, the recurring concerns are confidence (which we have covered at length, so I won't go into here), and the perception of what it takes to get yourself started. Notice I use the words 'perception of what it takes'

When it comes down to it, there is only one thing that will prove you have a service worth selling; Customers. If you have customers that will pay you for your service, you will make it.

And let's face it; people will always need plumbers. It's a job that will never become obsolete.

So long as we need water and a way to get rid of waste, we will always need plumbers.

To illustrate the simplicity of becoming your own boss I will use a plumbing scenario.

Imagine you are called to fit a bathroom from scratch. The room is a blank canvas and you have to decide where to place the bath, shower, WC and basin. There are factors that will be important and must be observed. Where is the soil pipe, doors, windows for example.

From here you draw a picture of where the fittings will go according to where the pipe runs will be. You set a picture of your end goal in mind before you start the work. From here, you can work backwards to establish which floor boards need to come up, what pipes to lay etc, etc.

Being your own boss is the complete opposite to this. It may be useful to think of yourself with a huge plumbing empire, employing many plumbers, and earning a mint. However, this may be very daunting and instead of acting as a motivator, may serve to make you feel that it's all too much, you may never get there and 'why am I even bothering'

The point is; *your goal needs only to be the next step ahead. Let me explain.*

My first goal was to get customers. When I had some, I modified my goal to having enough work every week to pay my bills. When I reached that goal I changed it again to having enough security of work to only work six days a week. That reached, I cut down to five then, I changed to making sure I could go away for a week's holiday and have work to come back to.

My goals now are that I can create an infrastructure for female plumbing students to enter, and help you to set up in business for yourself and a place online where customers can search for women plumbers anywhere in the UK.

Set some goals for yourself that you can reach. I can help by telling you that whatever you decide to do; you will be in demand. If you chose plumbing (I hope you do, but it's not compulsory) not only will I personally be promoting you through Stopcocks but your customers will promote you too.

As a woman you will be used to changing your goals. After all it's a woman's prerogative to change her mind! Some may think this is a negative trait or an insult. I see it as a necessary step towards meeting your aims. I urge you to see it this way too.

Research has shown that women don't make decisions lightly. They think deeply about all the possibilities before making up their minds. If we feel we've made a wrong decision we change our minds. My good friend Marcus De Maria is a stock market trader and teaches people how to successfully invest in the stock market. His research has shown that women are far more likely to make money trading, because they follow the rules and will buy and sell as necessary. Their decisions are based on the best outcome rather than the fact that they made the decision. He says his female students are a lot more successful than his male students:

> ***'Men get too caught up in their ego and will hang on to a stock hoping that it will come back up again, whereas a woman will sell it and move on'***

Yes, we change our minds, that is because we are flexible. Flexibility makes good business sense especially when the economic climate is volatile.

We have come on a long journey of reinvention to get where we are. We have been successful because of our ability to be flexible.

Think about your own life. Your children all have their own personalities, their own likes and dislikes. Your job has its own challenges. Your husband/partner needs you to be a certain way. If you were any more flexible you would be made of rubber.

Control your own destiny

Look at that statement and think about what it means to you. Your friend reading this same statement may well have a completely different understanding. This is where you dictate your own life journey.

Some may think being their own boss means working every minute of the day and earning as much money as is humanly possible. Some may consider it as successful to work three days a week, earn enough money to survive and be happy at that. You may want one goal for the first two years and change it to the other for the next two.

Destiny has no hard and fast rules. When you take it into your own hands there is absolutely no reason why you can't create it exactly as you please.

When I was at school all my friends had black onyx rings. I wanted one too. I saw one in the jewellery shop on the way home from school and every day I stopped and looked at it.

We were a poor family. We five children got 50p a week for pocket money (if we were lucky and the horses came in). The ring cost £4. I was determined, so I went without sweets for two months and saved my 50p until I was able to walk into the shop and buy the ring.

There was nothing special about the ring but the significance of what I had achieved was great. I decided I wanted the ring and I created a destiny for myself that I would have it.

I still have the ring in a box on my dresser. It's tarnished and worn thin with wear, and probably not worth a penny but to me it is a priceless reminder of what can be achieved when your mind is set.

All too often as women we are told we cannot do this or that, or we shouldn't dream. In a fit of rage one evening my father cried, 'What right have you got to ask for independence?' It was then that I knew I would have to create a new destiny. One that would be unprecedented in my family, and in my culture.

The chances are you will be far better equipped than I was. There is so much talk about business on the media. Women are able to get to the highest positions in the corporate and political realms. Although it is still a rare sight to see a woman plumbing, it is not unheard of.

There has never been a better time than now to decide your destiny and begin to create it.

Yes, you will have to be ultimately flexible. You will be your own secretary, bookkeeper, and labourer. Aren't you used to that already? Don't you already organise your family and your children? Don't you already bend over backwards to please all the people all the time?

Well this time, you can please yourself by pleasing other people too. Use those skills to your advantage, and the advantage of others.

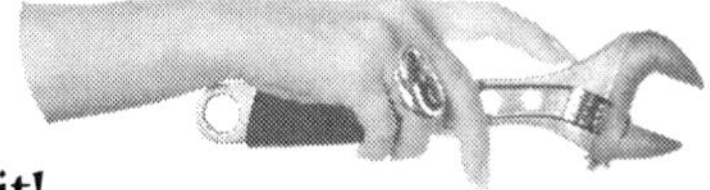

You're worth it!

If you have completed the exercises in the other chapters you will already have a compelling collection of documents showing you, what your dreams are, where your strengths are, and what your motivations are.

Take a new sheet and describe what your destiny will look like.

> *What will it feel like?*
>
> *What will it smell like?*
>
> *How will you know when you are there?*
>
> *Who will you be sharing it with?*
>
> *Where will you be living?*
>
> *What will you own?*
>
> *How much money will you be making?*
>
> *What will you do there?*

When I speak to people about female plumbers, many of them seem to have a story of some disaster they have had to do with a plumber or a builder, usually male. When I tell them I am writing a book they imagine it will be a book bashing tradesmen.

I spend a great deal of my time talking about opening opportunities for women in all the trades. I spend very little of my time talking about our male colleagues. Surely, there is enough room for both. I have often been accused of pitting women against men. This is far from what I am trying to do. I started my own business because I wanted to be my own boss. I did that early in my training because I couldn't get experience in the field in 1990. I receive many emails a week from women who are still in the same position I was in 20 years ago and I am in a unique position to do something about it. Promoting women in no way implies that I would pit them against men. To be frank; life's too short.

There are customers who don't think twice about the gender of their plumber, so long as she or he does a good job. The thing all customers have in common is that they want to trust the person coming into their home. They will be laying themselves open and paying money to have a job done and they should be treated with the respect that deserves.

It is unfortunate that some of our male counterparts do not take this into consideration, but that is in no way to say that male plumbers can't be just as respectful, and good at the job as women.

And so dear reader, we are about to step into the last part of our journey together.

We have come this far but as with any journey it changes and develops as we do. Before leaving you to take the next step into your own life, I have put together a recap chapter.

It will serve as a reminder of what you have achieved on this journey. Should you decide to go back and reread certain areas (and I recommend that you do), you will find references to each exercise here.

When I design a bathroom, I use a scaled-down version which I make out of squared paper. This way I can move everything around until I get the design right. In the real life bathroom, I then put everything in place and work out the order of installation. The whole job often requires me to revisit the drawing and enter new data (for example, pipe work I couldn't know about until I lifted the floorboards). This book is like that.

You will come across things that make you want to revisit some of the chapters again.

If you are the type that likes to keep your books in a pristine condition, then I recommend you buy two copies.

Web links

If you want to hear some stories from successful women plumbers just go to www.joyofplumbing.co.uk/downloads or to join our online community or simply to find a woman plumber in your area go to **www.stopcocks.co.uk**

By going to **www.joyofplumbing.co.uk/downloads** you can get the audios and worksheets, to do the exercises as often as you need.

CHAPTER NINE

Your new life begins here

This is where our journey ends

We've come a long way. You've come a long way. If you are the type of person that likes to read a book from the end then this section will be very useful for you. Our journey together ends here but your own journey continues.

I'm about to give you a potted version of the journey so far. You'll also find links to everything at the end of this chapter.

In the introduction you decided this was a journey of self discovery you wanted to embark upon. We began in Chapter 1 where we looked at the art of reinvention. Women reinvent themselves many times in their lives to fit in with family, friends, partners, community and society. We discussed reinvention as a necessary part of our lives and saw that if you make a list of all the different ways your life has manifested you will realise that reinventing yourself as a plumber is just another direction you can go in. I invited you to take a long, hard, honest look at your current life and gave you a technique to get reconnected with your dream. Together we banished guilt and got permission to speak frankly about how we feel now and what we want to change.

The next part of the journey took you deeper into your own life and used the Fulfillment Circle to assess just how happy you are with your life and your job.

This is where you saw for yourself, your life in three colours. The honest truth about how fulfilled you are in your life and work. You were invited to think about the intended legacy you wanted to leave and I shared mine.

In Chapter 3 I told you to get out of your own way and let success happen. The process of reinvention has a number of stages. We walked through them. Starting with another long hard look at where you are, what you feel, why you want/need/must change your life. You did some deep soul-searching and looked at the directions you wanted your life to go in.

We asked the big question, 'Why do women have confidence issues?' and looked at the expectations of others, family, friends, partners, society to try to unravel the difficulties we as women have in striking out for what we want.

I challenged you to allow your new life in and I shared with you the process I went through including the 'warts-and-all' way I put off making my decision to step out and take control of my life.

By Chapter 4 we were looking at the progression in history of women in the trades. With the kind permission of Professor Joanna Bourke we saw that during wartime it was perfectly fine and indeed expected that women would go into the trades to build the machines of war. That 'let the cat out of the bag' or should I say 'the women out of the kitchen'. The struggle for equality rages still and as women plumbers we are still a novelty. I suggested the number of women in these roles would never equal the numbers of men in them and I rejoiced in that. As long as we are news and we always will be, we can be sure of success. The point was to see that as a strength rather than a weakness. The fact that even after 20 years as a plumber I have seen very slow progress has been both a blessing and a challenge. Accepting that most women won't want to be plumbers, (neither will most men, for that matter) has enabled me to concentrate on you and your desire to become a woman, successful in a man's world.

In Chapter 5 we explored why customers love female plumbers and why they will love you. I highlighted three of the 39 top complaints customers have about tradesmen (and they were mostly all men). The actual report can be read in part at the end of this book. I speak to plumbers both male and female all around the UK about how to improve their businesses. In one instance I suggested the guys wear a name badge with a photo on. This made an immediate improvement to their customer relations.

Here was our first look at the archetypes. Characters and characteristics of our behaviour that tap into the deeper meanings of our relationships with the world around us. The emotional connection we as women have with our customers is our strength. This creates a trusting relationship.

Believe it now. **Customers love female plumbers.**

Having recognised that customers will love us, we moved on to looking at our physicality. Every woman knows and feels on a daily basis the pressures of having (or more often, not having) the perfect body. Far from just looking at the female form, although it was looked at and deemed flexible enough to get into all the nooks and crannies plumbing requires, we also delved into our strengths as women. We discovered that being generally smaller in bulk means that our strategies for doing certain aspects of the job are more a case of using our brain than our brawn. Not only are we carrying the weight of two baby elephants all the time but then we find our strength is made up of all our trials and tribulations from our entire lives, things that have been piling onto us that we have simply had to get used to. We heard the amazing story of my tiny, aged mother in law who was able to tap into her hidden superhuman strength to save the life of a young boy, showing us that we all have it within ourselves to be as strong as we need. You took another look at your life and drew strength from all the challenges you've already overcome and I shared some of my own.

In the seventh chapter we got down to basics and looked at ourselves in comparison to men. Far from this chapter being about how we are better than them or they are better than us, we were able to see how we have complemented each other since pre history. In my observations of male behaviour around the world I wondered if the male role as protector and hunter had been usurped and left him unable to fill his time. (Or if he is simply an idle creature, like a male lion).

For years the struggles have raged about 'equality'. We looked at the meaning of 'equality' and applied it to ourselves. Does being equal mean we have to be like men and they have to be like us?

I challenged the term 'equal opportunities' and pointed out that being treated the 'same' was not equal at all. I challenged also, the notion that inequality is something done to us. We again saw our old friend confidence come to play. We acknowledged that the developed world was extremely lucky to have the luxury of time to think about ourselves in the way we do and I invited you to see some inspirational footage.

We discussed that neither family, friends, the law nor society were to blame for our own feelings of inequality and that equality was a state of mind that you could achieve by taking a first bold step. We learned that being lucky meant we could be grateful and that gratitude leads to boldness and being bold means we can begin to create our own luck. I invited you to become 'equal' in your mind-set and make it come true in your life.

Then we explored yet deeper into the archetypes that embody universal human traits. We saw that by slipping effortlessly between the male and female archetypal roles could make us appear superhuman, that doing this touches the hearts of our customers in ways we can't possibly truly understand but can benefit from enormously in our lives and our business.

We saw that male and female archetypes explain our differing approaches. The feminine approach is multi-dimensional seeing person, problem and solution, whereas the masculine approach is to see the problem and the solution. Both are very necessary to do a good job.

In Chapter 8 it was time to look to the future. The comparison between the old ways of running a business and the ways of new business showed us what the new order is. Gone are the days of needing tons of cash to start up. Now all you need is a phone, a laptop with internet access and a few second hand tools. Big business is too slow to keep up with the changes in modern society. Now it suits us to be as flexible as we can and no one is more flexible than you. We noticed that running our own business is as simple or as complicated as we make it. We heard examples of women being ousted from jobs that couldn't bend to suit their lives.

You created a destiny for yourself, you determined what success means to you and you asked yourself what does it mean (to me) to be my own boss and noticed that it doesn't have to be dictated by anybody else but you.

Then you dared to dream the dream and saw what your own destiny will look like. You went back over all the exercises of previous chapters and saw that you had a catalogue of your life, your achievements, and you had identified your future.

So here we are at the penultimate chapter and I am ready to set you off on your journey beyond here.

I will not be leaving you to fly solo for evermore. In the next chapter, some of my friends and mentors have given their expertise to give you further strength.

Web links

See the Stopcocks Channel at **www.youtube.com/stopcocks**

Get all the downloads, from **www.joyofplumbing.co.uk.**

If you are already a student plumber, woman in the trades, or established female plumber or if you would like to contact me you can join our global community at **www.stopcocks.co.uk**

Embrace your inner girl can be seen here **http://www.ted.com/talks/lang/eng/eve_ensler_embrace_your_inner_girl.html.**

Good Luck!

CHAPTER TEN

What Others Say

Don't Take My World For It

In writing this book I have learned so much and so many of my friends and mentors have been kind enough to contribute their knowledge and expertise.

Enjoy reading what experts have to say about the very personal topics I have been covering in creating this book.

To Be Strong Is To Be Feminine

Goddess Star Monroe

As they say it's a mans world but as the late, great James Brown sang with such passion 'it ain't nothing without a woman or a girl'. Women! Wow what fabulously magnificent creatures we are.

Women are an integral part to this wonderful world we live in and within each woman there is a powerful force filled with amazing instincts, passion, creativity and ageless knowing.

Women explore, discover, nurture and inspire. Women have a softness, a gentleness, an expansiveness that is innate to them. It enables us to yield, flow, open, endure and connect. These are all parts of our feminine superpowers.

When you go deeper you discover that at the core of every woman is her genuine inner strength. It's a strength that is true to her, it courses through her veins and arteries, it charges her up and sets her on fire. It's a strength thats been passed down from generation to generation.

Women's strength is different to that of men, ours is deeper, stronger, it has roots and it has connection, a men's is more abrupt, competitive and sometimes can be aggressive. Our strength is clever, its wise, its exquisite. It comes when we embrace our softness, our gentleness and our expansiveness. Woman's strength has a grace and elegance that weaves and charms its way through enabling us to endure and stay grounded and be true to who we really are.

A woman who is in touch with her real self is beautiful, happy, flowing, she enjoys life and is full of a unique energy that is contagious to everyone around her. This woman knows she has this deep rooted strength, she is at one with it, she trusts it and she uses it wisely to move her forward through her life. Her strength is her knowledge, it makes her wise, patient and loving. She knows she will meet with challenges and obstacles along the way but trusts and uses her instincts and intuition. When this woman entwines her strength with her receptive and magnetic qualities of openness and gentleness this woman is completely living in her feminine side.

Women need not shy away from being and feeling strong, as you see it is an integral part of being a true woman. Strength is not at the expensive of femininity but more an enhancement of it. Women who know this shine in all their power, grace, strength and beauty, they are mesmerizing, enchanting and captivating, their energy is magnetic and their sense of possibility is contagious.

Personal message from Star Monroe

I have had a very eventful life. Amongst other things I have gone through a divorce, struggled with various eating disorders, have had long dalliances with alcohol and drugs but all the time I was rocking my life out at the bottom I deeply knew that everything would be ok, that I would survive, that I would move forward when the time was right. It was this inner self belief that was wrapped with my innate womanly strength that helped me move forward with my life. I can honestly say that I am a normal woman, a real woman, a woman that has felt so much hurt, sadness, guilt and fear but I tapped into my feminine superpowers, I clung on when times were tough and 2½ years later I am living proof that inner strength and belief can conquer all. I am clean, no longer have any issues surrounding food, I am a great mummy to my son, a fabulously passionate and successful entrepreneur and I'm absolutely in love with me and my life.

I am a modern day Goddess and a crusader of the truth for women. I am a force, an unstoppable woman. Extraordinary, extravagant, glamorous, tough talking but with a kind heart. I am wise, passionate and I love being a woman. Through my spirited, fun and sexy teachings I empower and educate women everywhere to rock their most awesome lives and bodies! Its all about the Goddess Star Monroe lifestyle.

www.goddessstarmonroe.com 

Making It All Work

Bridget Biggar

One of the major contributors to creating resilience is finding ways to be happy. Some UK research carried out in 2010 showed that people generally consider women to be more resilient than men, however in many surveys women believe they are less happy than they were in previous generations. So where have we lost this connection? One of the reasons is that there are far more opportunities for women to succeed in professions which were never before dreamt possible. This does mean that we now have a larger group of people to compare our successes and failures with, as well as having increased expectations from our jobs. Younger women in particular are attaching greater importance to the significance of 'success at work', 'being a leader in their community' and 'contributing to society'. Having these increased expectations can work to our advantage though – particularly if we are in a role that gives direct and tangible service to others – such as being a trusted, skilled provider of solutions and increasing people's standard of living. It is often the way we think about what we do that actually affects how we feel.

The good news is that you can easily find ways of increasing your job satisfaction and happiness. You can boost your happiness by doing what you choose and liking what you do.

Thinking about your choice for pursuing this career first, often we spend far too long thinking rationally about complex problems. Surprisingly it doesn't necessarily give you the best answers. I bet that "common sense" is high on any plumber's list of strengths, so how about trusting gut instinct rather than "thinking through the pros and cons"? Is it that because you might be moving into a non-traditional role, you feel it needs to be justified with reasons when actually you just want to do it? Simple problems require reason; novel ones need creativity to apply hunches and the ability to tap into past experiences that give you clues as to how you successfully solved unexpected issues.

Strong decision making tactics when deciding on a career change, for example, might be

'Will I regret not having tried this?'

'Will I be able to use my strengths frequently?'

'Will the things that are important to me (eg being my own boss/ working at certain times/meeting new people/creating beautiful environments) be part of everyday life if I make this move?'

Then tap into your intuition and gut feel to answer them.

Asking yourself the right questions when making momentous choices ensures that the answers you give will create a resilient basis for the future. If you decide on a career change for the wrong reasons, say, because of the potential to make more money, without factoring in that you hate meeting people or working in the evenings, you will rapidly feel unsatisfied and you won't stick it. Doing something for the right reasons is critical.

The second key element of resiliency is the ability to identify and play to your strengths as much as you possibly can. To do this, think through experiences that really stand out as high points. What were you doing? What does that tell you about your strengths? There tends to be a very strong link between us using our strengths and feeling satisfied and happy. Not only will identifying which strengths you most like using give you a clue as to what you should be spending most of your waking time doing, but can tell you what it is about you that you should be marketing. This will attract colleagues and customers who feel drawn to your particular strengths.

We buy from people we like and feel comfortable with.

So if you get a real sense of satisfaction from doing a thorough, detailed, systematic job, then advertising this will attract those customers who want to know every detail, will have a clear plan and are probably very cost-conscious.

Maybe what really excites you is using your creativity and artistic abilities to make something beautiful to look at that wasn't there before. Perhaps your skills at finding unusual combinations and getting people excited by "what could be" would really bring you joy.

Using a natural ability to do clever things with your hands rather than with words is so envied and brings such a sense of personal satisfaction, that you will never look back.

Sometimes meeting new people and really feeling that you are making a difference to them might be the thing that makes you look forward to the week ahead.

Or even the sense of achievement every time you come up with a solution that makes a big difference, and perhaps getting the job done in record time.

It is really important to realise that when we are nervous, tired or in unfamiliar situations we can overdo our strengths and then not get the result we are looking for. When threatened we might become pedantic, wacky, idealistic or bossy – or at least come across like that to people who aren't like us. So recognising this can really keep us satisfied and sane.

Finally, we can get in the way of what we can achieve with the way we talk to ourselves. This tends to be the way we describe to ourselves what's happening. We can really undermine how we feel by believing, for example, that we have no control over events. Imagine how de-motivational this is. It doesn't have to be true or false because it's our own perception that dictates how we approach every set-back or opportunity. De-personalising issues is also vital. Things don't necessary happen 'because it's something wrong with me or I'm a woman'. Neither are they permanent – 'it's never going to go away' 'it's always going to be like this'. And also whatever happens, good or bad, is not a predictor of everything else that might happen in any other part of your life.

Negative explanations can be such a part of your habit that you might not believe that things could be described any other way. But the next time something doesn't go right, pay attention to how you describe it to yourself and check whether it's optimistic or pessimistic. Realistic optimism doesn't mean denying unpleasant facts. It means differentiating between facts and your interpretation of them. Choose to think about things in a way that is helpful to you.

Lastly, make sure you pay attention to the positive things that happen as well as the negative. If you can make yourself feel good you will be more resilient and resourceful to solve the things that maybe don't go so well, or when you face adversity or set-back. You must retain your belief in your strengths and personal competence. Focus on the experience and competence that you have used in the past and know that you can use those things again.

And – people matter. We need others to make us feel connected, loved and stimulated. People want to help and they will most help (and buy from) those who exude enthusiasm for what they do well and enjoy. Take yourself seriously. Contrary to popular opinion, people like winners. If you have a belief in yourself, then so will others.

Bridget Biggar

Director of Talent for All About Brands plc, and Exclusive Agent for the Lifo® Method in the UK, Ireland and the Middle East, Bridget has a Masters in Applied Positive Psychology from the University of Pennsylvania. She works with individuals and organisations of any size, throughout the world to help them improve their performance through the identification and application of strengths and innovative joined up thinking on personal performance and flourishing.

You can find out about All ABout Brands here

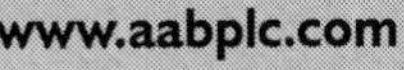

The Feminine Way of Being

Susie Heath

'Greased nipples, ballcocks, lubricants, tools, bits, plungers and male and female unions' always make me laugh. As terminology for a plumber, this is every day language, but for those of us wanting our central heating fixed and our boilers serviced, it comes somewhat as a surprise that we have sexist vocabulary for bits of metal pipes and tubing! Spoken by a man, it can come across as just a little bit risqué for a female client as we are asked to 'turn up the heat,' whereas something to share a giggle over with a female plumber.

Interesting isn't it that during the Second World War, women became proficient in what had been typical male trades, yet now there are still raised eyebrows when women choose such trades as their profession? You, like me, realise that women can do just as much as men, and very frequently more and (please don't tell them this) often in a better way. Our only drawback is that although we are physically strong, that strength is not quite as sustainable as a man's over a long period of time.

Sadly what has happened in this still male-dominated world, is that women have tried not only to fight men at their own game, but to overtake them - and it has worked. In this search for acceptance, women have tended to over-adapt towards their masculine side. We have fought with men in courts, in board rooms, on building sites and in war zones to ensure that we have been seen, heard and acknowledged for our courage, strength and endurance. But our bodies have not been designed for such endurance, and the battling has taken its toll in such a subtle way, creeping up on us in ways which are unimaginable.

We have adopted male attributes and attitudes towards work, taking on masculine body language, male talk, voice tone and stance, and even male clothing to fit the job, while our femaleness flies out the window. Thus deprived of femininity and gentleness as our new male responsibility adds stress, strain and worry to our lives, our serenity vanishes and eventually our families suffer. But take heart – there is a better and more rewarding way.

Am I suggesting that you cannot be a scientist, a mathematician, a train driver, an engineer, an inventor, a lawyer, a scientist, a bricky or a plumber if you're a woman? Not one iota. What I am suggesting is we be all that and more, but in our way, in the way of woman. Do what you do in a feminine way, not by taking on the attributes of a man, not by competing with him. By giving all that you do the feminine touch, you can enrich society in a beautiful way.

Please don't be cajoled into thinking that femininity translates to weakness. Our feminine side is a force that can conquer all - (just look at an animal protecting her young!) There are times when we have to be strong and forceful, creative and energetic, powerful and courageous - this is our Yang energy, our masculine side which we need in order to take our work out into the world, for making important business decisions or pushing our babies out of the womb. But too much Yang creates havoc with our hormones and menstrual cycle, with our intimate relationships, our fertility, and also with holding on to pregnancies. Losing ourselves in an imitation of men makes us deeply unhappy in the long run. Yang energy is our emergency energy – we are not supposed to live there! When we continually struggle effortfully to adapt to an energy which is not who we are in essence, we create so much pain for ourselves as we fail to honour the deep feminine inside.

Our Yin energy – our female side, needs to be our main energy. This doesn't mean being pink and fluffy, or just wearing sassy underwear and having our nails done. There is far more to us than that. When we can learn to appreciate our own honesty, dependability, kindness and love, our sweetness, our gentleness, our prowess, our passion, our sensuality and our joyful eroticism, we will attract into our lives just the sort of people who will directly complement and support those energies. And yes, we can still do and be this as powerful successful businesswomen!

Unless we allow our feminine to be truly seen, we are in danger of a whole generation of girls missing out on the joys of real womanhood and our true nature. Laddish culture means the innocence, elegance and joy of femininity is lost to them, and the whole world becomes poorer for that.

My plumber happens to be a young woman who wears a pink boiler suit, and we have swapped many girly stories over 'vibrations in my pipework.' What makes her so special is the fact that she does her work with total enthusiasm, excited that she is able to achieve her outcomes, and full of confidence that she will find a solution – self-assured enough to know her limitations and being willing to ask for help when needed. She is a gem and much appreciated by all who work with her.

Perhaps now we have proved ourselves, we can stop striving so hard, give in gracefully at last to our natural instincts and come back once more into our soft, yet powerful, loving feminine. You can stop competing with men and learn to be your true loving feminine self in whatever it is you do, because it is then that your light will shine – clients will come to you in abundance because your energy is so appealing and attractive. Women tend to be far better communicators, more compassionate, and more understanding than men – it's in our nature. We tend to want to improve things, to leave places looking even better than when we arrived. It's not a weakness this, it's our strength.

How would it be if you now claim and own your profession in a feminine way, rather than just emulating men? Realise that you have a wonderful role to play as a woman in your own right, and enjoy every single moment.

Susie Heath

Relationship and Intimacy Coach, Life Strategist, Business Coach, Biodanza™ teacher and Author of "*The Essence of Womanhood – re-awakening the authentic feminine.*"

Go to **www.essenceofwomanhood.com** for a FREE audio relaxation download

Alignment and Balance

Mica May

Many years ago I was a dancer.

I quickly learned that if your alignment isn't spot on your balance fails and you fall. This hurts; sometimes a lot.

I have carried that lesson with me throughout my life and though sometimes I have forgotten it for a while, I always re-remember it when I fall.

My suspicion now is that it's a Universal Truth

In many years of working with clients one of the things that has struck me is what we all have in common, that, ultimately what my clients want is to have a happy life; how we get there is different for each of us, but when it comes down to it we want the same thing. I'm talking about here in the West where we're living a life largely separated from the natural environment and close to people we haven't known all our lives and the flow of the seasons, which tend to impose a certain structure.

Bringing the lessons of my life as a dancer into my work as a coach, I am aware that for us to to achieve that happy life we need it to be balanced and for that we need to be able to align what we're doing day to day to our core values.

Doing that seems after all these years to be probably the most important task we have to complete to achieve this thing we refer to as 'work/life balance'.

I argue with that term a bit, it makes it sound as though when we're working we're not living and vice versa. Actually, all of us, but especially women, are forced to integrate our work into our life and our life into our work constantly.

We fit the bit for ourselves, that part we refer to as 'life' around the part we call 'work' (usually what we do to bring in an income) but often also, the housework, shopping, looking after our children and partner and meeting our own needs for satisfaction if we have time when everything else is done.

Most of us work to live and also at the same time, we live to work, or hope to.

Even if we don't especially enjoy the actual job we do, not having a job or work of some kind is not good for us, considerably worse in fact; and not just financially.

If we have no work not only do we amass debts or have to tighten our belts to a ridiculous degree but our self esteem plummets, we lose our sense of routine and generally it's easy for things to begin to fall apart. We don't even have to fall victim to any of these ills, for the stress of them attempting to creep towards us to have negative effects on our lives. Although at the same time there is a fantasy we hang on to that being in that situation would give us time to live our dream, the people who manage that are very rare indeed.

There is plenty of evidence that tells us that if we love what we're doing for pay (whether this is a job or if we're working for ourselves) we will be better at it and more fulfilled in the other parts of our life; also, we are more likely to be successful in all areas of our lives. And because we're happy, we're motivated, so likely to do well and get well paid, though that is less important to us if we're living a fulfilling life anyhow.

So what about core values? How do we know what they are and align our life with them?

It's a fact that when aspects of our lives are out of alignment with our core values we suffer from stress and worse. We can cope if some parts are out or we're out a little, but the worse it gets the worse is the resulting stress.

These values are what really motivate and inspire us and enable us to create, create change and to have real, lasting determination.

It may for example be one of your core values to be trustworthy or perhaps you're a perfectionist. Being surrounded by positive relationships may be a necessity for you which can make working in isolation difficult. You may value order or working with your hands or not having other people tell you what to do.

Firstly, remember that these values can change. It's important not to simply stick to something because you wanted it once and your childhood dream of being a fireman may or my not be aligned to your values now you're an adult; but lets unpick the fireman fantasy a little.

Whether we were girls or boys, many of us had fantasies as children of becoming someone heroic. Perhaps you have been able to translate that into something in which you are able to help other people on a daily basis. Although you may not be galloping over the hill on a white charger (mixing my metaphors), but your job as a fantastic accountant makes you a hero in some people's lives!

Or perhaps not, maybe what you're doing is completely unsatisfying, you need to be more practical too, to save people and situations more directly? Maybe, as I suggested before, you don't like other people telling you what to do and are great at problem solving.

So, for you one of your core values is likely to be that you can have a positive impact on the people you encounter daily, and another that you need to be in an environment without a rigid hierarchy and if you're not getting that, your life will feel very unbalanced.

In the hunt for your value take a look around your house; what do the objects you surround yourself say about what's important to you? Who are the people you choose to have in your life (do you even really like these objects and people)?

Sometimes we do this and realise we've settled for a life that isn't fulfilling, we struggle with discontent and our two weeks holiday and too short weekends just don't give us enough.

Don't take this out on your partner, take a good hard look at yourself and discover what you would love to be doing.

You don't have to change all at once but you can begin and begin now to take small steps towards that life on a daily basis.

In this case, size really doesn't matter, but taking those baby steps does!

Hattie asks, as do I, since most of us reinvent ourselves constantly all through our lives, why not do it in the direction of what we'd love to do rather than what it's convenient for us to do; rather than what makes other people's lives easier?

Yes, it may appear selfish, but if we're fulfilled then we will be happier, healthier and even, wealthier; and isn't that going to be of benefit not only to us but to everyone else around us far more than us sacrificing ourselves on some imagined altar of 'duty'?

And if we have a great dream of doing good, of benefiting the planet, having our own business, it's important we don't wait, or lose that dream to cynicism or lack of energy but that we begin taking those baby steps towards it today.

Many of us have goals in life. Things we aim towards; sometimes those goals like our dreams and values are what we wanted years ago and have changed now.

It's fine to change your goals. There is no benefit in clinging to something you no longer want.

But if achieving them has any importance to you at all, make sure they are aligned to those core values and you'll find everything neatly fitting together; becoming smoother.

Perhaps you have a big beautiful dream you're too shy to tell anyone about because you're surrounded by other people all not saying either? Maybe the world is full of women wanting to be plumbers but too afraid of being laughed out of the room to say anything about it...

I'm not saying following your dream and aligning your goals with your values will mean you will leap out of bed every single morning and dash energetically into the day ahead, eyes shining and nose all wet, but a balanced life in which you know that every day is taking you closer to something you hold very dear is a truly fulfilling life; and once you have begun to get near, and it doesn't have to take long if that dream is your own plumbing business you really will begin to feel like that the majority of the time.

When all aspects of your life are aligned, this ease can happen and you'll wonder why you didn't begin sooner.

When you're clear about what you're working towards in your whole life and are consistently moving into that, not dragging yourself through each day waiting for the weekend, the next holiday or retirement; when you can see, feel and smell your goals getting close and everything you do brings them nearer, the whole world becomes different.

Even things you hate feel less of a drag because you are doing them as a step towards that shiny, beautiful life you want.

So, if you, like Hattie want to change the way the nation feels about tradespeople, go for it, if you have a thing about copper or water; become a plumber. If you want nothing of the sort but you're discontented with your life, begin moving towards what you do want, maybe just as a first step, identify it but become inspired and feel the wonderful quality your life will take on.

What do you have to lose?

Imagine if everyone in the world was able to work towards creating something they loved or to do on a daily basis work that brought them satisfaction, how different would everything be?

Having a balanced life may be about having a life that means you can have long wonderful holidays in the sun, having your own business or volunteering in Africa.

Discover what rocks your boat and move everything around in order to make it possible.

Start the revolution; ask yourself what you'd love to be doing in your life and make a step towards it.

Ask yourself if everything you're doing points in different directions and throw out the distractions so you can begin following your star, allow your life to fit together perfectly.

Ask yourself, why you aren't doing that already and be really strict about the answers you'll accept; no flim flam allowed.

What changes can you begin to make every day, however small, that will bring you closer?

Start making them.

Mica May

Coaches business people towards greater satisfaction through improving the alignment in their lives. Using her work as a play specialist she combines this with her expertise as a coach with professionals to enable clients to bring more joy into their lives and now extends that to families and individuals with her play sessions.

mica@micamay.com

Other Voices

APPENDIX

Here is an extract from Andrew Priestley's Study. The full report can be obtained by going to www.joyofplumbing.co.uk/downloads

If you'd like to spend 1 ½ hours with me chatting about applying any of these ideas in your business call **07545 316451** now.

I consult with plumbers UK wide about improving their businesses and this service. Knowing what the customer is worried about can help you to provide a service that will out perform the competition every time.

You can book a £125 session in which we go through the report in detail, where we identify areas to improve your business and strategies for implementation.

Special Report: The most common and potentially costly mistakes made by tradespeople

In this special report by Andrew Priestley asked 317 customers to retell their worst 'trades industry' experiences. The report provides feedback to business owners and staff about adverse behaviours that actually impact on a customers decision to engage or reuse the tradesperson.

Introduction

There is a theory that:

1) that if you **ask** your clients to identify the key frustrations they have with you, your business or your industry… they will usually tell you;
2) this feedback constitutes a reliable to-do list of things to **fix** in your business that your clients currently don't appreciate;
3) that addressing these issues can result in new and repeat business and referrals.

Interviews were completed for a client who operated a home improvement centre; and a centralized booking service for tradespeople - trades, sub contractors, service providers, to-site and on-site suppliers.

The participants

We interviewed 317 customers - new, existing and lapsed customers from the database of the home improvement centre. The participants were asked open-ended questions about their **concerns, frustrations and complaints** with:

1) any recent specific experience with a service provider
2) the 'service industry' in general – carpet cleaners, tradesmen, bricklayers, carpenters, fix-it men, plumbers, dog washers, pest controllers – basically anyone who provides services in the home; and

Tip:

Your customers are a source of valuable information about you and your business but you have to ask them. Every business should survey their customers from time to time. For our purposes just regularly asking all your customers for feedback offers immense benefits.

You can ask customers to complete post service delivery questionnaires or have someone phone or visit a customer for about a short chat. We tape-recorded the interviews so we could collate the feedback, transcribe the data and use it as a training tool.

The participants

Customers were specifically asked to focus on their **concerns or complaints. And there was no prompting.** (i.e. making suggestions, giving them examples to get them started.)

We determined:

- the service being provided
- the critical incident or incidents
- the gender of the tradesperson (i.e. male or female)
- perceived age group
- perceived level of expertise
- dissatisfaction rating on a scale of 0-5 ; 0 being Very Dissatisfied through to 5 being Very Satisfied.

The comments were then tabulated on an Excel spreadsheet.

While there were many complaints, they clustered down to 39 key issues. This report lists, in rank order, **the 39 most mentioned complaints** e.g., 1 scored the most points and so on.

Four Phases of the Transaction

As we discovered, the customers are telling us what upset them across four phases of the total transaction:

1) canvassing/lead generation
2) quoting for business
3) service delivery
4) post service delivery complaints and follow up

They then provide the key reasons why the tradesperson didn't get the work. We note that often this is irrespective of the competitiveness of the quote, expertise or availability.

The report provides a general complaint (i.e., He didn't show up on time.) then an explanation and a recommendation.

Note: The word 'he' is used throughout this report and sadly, in most cases the interviewee is actually referring to a **male** *service provider. This* *point was cross-referenced VERY carefully. There were instances where the complaint was about a female service provider, but these were few.*

Tip:

If you are MALE, read this very carefully because it probably applies to YOU … exactly.

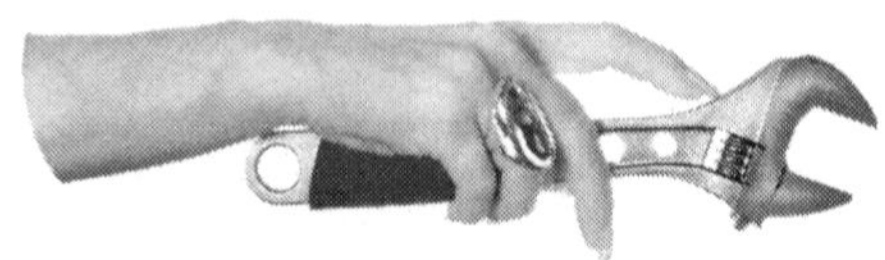

To get the full report and book a 1 ½hr session with me go to **www.joyofplumbing.co.uk**

or call **07545 316451**

Hattie Hasan was born in North East London to first generation Turkish Cypriot parents. Rather than enter an arranged marriage, she studied Psychology at London University and went on to attain a post graduate degree at Reading University. She taught in several inner London schools until 1990 when she retrained as a plumber and started one of the first companies of women plumbers in the UK. Stopcocks Women Plumbers now encompasses self employed female plumbers throughout that country. Hattie travels speaking to schools, colleges and industry about the importance of women plumbers and supports female students to start their own business.

She lives in rural Yorkshire with her partner and dog

Testimonials

"A charming book, packed with useful advice and real insights. Useful not just for would-be women plumbers, but anyone looking to make a difference to their lives"

Mike Southon

Financial Times Columnist and best-selling business author

"The Joy of Plumbing is a wonderful, irresistible book: perceptive, funny, compassionate and always breathtakingly honest. Hattie Hasan invites us on a compelling, personal journey of discovery, celebrating the strength of women with warmth, wisdom - and water. A lasting liquid legacy!"

Mike Symes

MD Strand Financial

"It's a really clever blend of fact, memoir and motivational treatise, about personal development and the power of change and that can be applied to anyone and I really loved that about it - it's very well put together."

Dr Rachel Connor

Author, creative writing facilitator, lecturer

"I love the way you break everything down. It makes me think anything is possible. It makes me want to fly back to Africa and do what I love doing. It's very inspiring to young women. "

Huriye Solkanat

20yr old student.

Printed by Amazon Italia Logistica S.r.l.
Torrazza Piemonte (TO), Italy